UNDERSTANDING ELECTRIC POWER SYSTEMS

IEEE Press
445 Hoes Lane
Piscataway, NJ 08854

Technical Reviewer
Bruce F. Wollenberg, *University of Minnesota*

Books of Related Interest from the IEEE Press

Electric Power Systems: Analysis and Control
Fabio Saccomanno
2003 Hardcover 728pp 0-471-23439-7

Understanding Power Quality Problems: Voltage Sags and Interruptions
Math H. J. Bollen
2000 Hardcover 576pp 0-7803-4713-7

Principles of Electric Machines with Power Electronic Applications, Second Edition
M. E. El-Hawary
2002 Hardcover 496pp 0-471-20812-4

Analysis of Electric Machinery and Drive Systems, Second Edition
Paul C. Krause, Oleg Wasynczuk, and Scott D. Sudhoff
2002 Hardcover 624pp 0-471-14326-X

UNDERSTANDING ELECTRIC POWER SYSTEMS

An Overview of the Technology and the Marketplace

Jack Casazza
Frank Delea

IEEE Press Understanding Science & Technology Series

IEEE PRESS

A John Wiley & Sons, Inc., Publication

Published by John Wiley & Sons, Inc., Hoboken, New Jersey.
Published simultaneously in Canada.

Library of Congress Cataloging-in-Publication Data is available.

ISBN 0-471-44652-1

10 9 8 7 6 5 4 3 2

Our thanks go to many who helped with this book but particularly our wives, Madeline and Irene, who provided support and encouragement.

Jack Casazza
Frank Delea

CONTENTS

LIST OF FIGURES

PREFACE

As Joseph Swidler, former Chairman of the Federal Power Commission (predecessor of FERC) often stated, "There are many disagreements about the best electric power policy for the USA, but there is no disagreement it is often being established without adequate analyses." Government and business decisions on electricity supplies often fail to recognize how power systems work and the uncertainties involved. Those involved do not always mean the same thing although they use identical words. Incorrect assumptions have been made about the operation of the electric system and continue to be made based on the operation of telephone systems, gas systems, and other physical systems that are not applicable to electric power systems.

The purpose of this book is to help those in government, business, educational institutions, and the general public have a better understanding of electric power systems, institutions, and the electric power business. The first nine chapters focus on the technology of electric power; the last eight cover the institutions and business practices. Why must business practices be included in such a text? Because technical and institutional practices need to be coordinated to meet our needs. New technologies require new institutional approaches; new institutional mechanisms require new technology. Both must be understood.

The original text for this book was written in 1984. It was used for instructional purposes in a number of courses for electrical engineers who were not power systems engineers, for lawyers, accountants, economists, government officials, and public interest groups. Since then some technological changes and many institutional changes have occurred. With the advent of the internet, many new and valuable publications and information sources have become available and were used in its preparation. It includes ideas and information from many segments of the industry and many knowledgeable people in the industry, and is based on educational programs of the American Education Institute (AEI).

The book covers such subjects as electric power systems, their components (generation, transmission, distribution), electricity use, electric system operation, control and planning, power system reliability, government regulation, utility rate making, and financial considerations. It describes the "six networks": (1) the physical network, (2) the fuel/energy network, (3) the money network, (4) the information, communication, and control network, (5) the regulatory network, and (6) the business network, which are interconnected in the provision of electric power. It provides the reader with an understanding of the equipment involved in providing electric power, the functioning of the electric power system, the factors determining the reliability of service, the factors involved in determining the costs of electric power, and many other technical subjects. It provides the engineer with background on the institutions under which power systems function. It can be used as a classroom text, as well as a reference for consultation. While a book of this length cannot provide in-depth discussions of many key factors, it is hoped it provides the broad understanding that is needed. Ample references are provided for those who wish to pursue important points further. The index facilitates the location of background material as needed. The authors welcome comments, suggestions, additional information and corrections. They hope you, your company, and all consumers benefit from it.

Jack Casazza
Jackcasazza@aol.com

Frank Delea
Fdelea@aol.com

American Education Institute
www.ameredinst.org

1

HISTORY OF ELECTRIC POWER INDUSTRY

ORIGIN OF THE INDUSTRY

The electric utility industry can trace its beginnings to the early 1880s. During that period several companies were formed and installed water-power driven generation for the operation of arc lights for street lighting; the first real application for electricity in the United States. In 1882 Thomas Edison placed into operation the historic Pearl Street steam-electric plant and the pioneer direct current distribution system, by which electricity was supplied to the business offices of downtown New York. By the end of 1882, Edison's company was serving 500 customers that were using more than 10,000 electric lamps.

Satisfied with the financial and technical results of the New York City operation, licenses were issued by Edison to local businessmen in various communities to organize and operate electric lighting companies.[1] By 1884 twenty companies were scattered in communities in Massachusetts, Pennsylvania, and Ohio; in 1885, 31; in 1886 48; and in 1887 62. These companies furnished energy for lighting incandescent lamps, and all operated under Edison patents.

Two other achievements occurred in 1882: a water-wheel-driven generator was installed in Appleton, Wisconsin; the first transmission line was built in Germany to operate at 2400 volts direct current over a distance of 37 miles

[1] Homer M. Rustelbakke, 1983, *Electric Utility Systems and Practices, Fourth Edition*, Wiley, New York.

Understanding Electric Power Systems: An Overview of the Technology and the Marketplace, by Jack Casazza and Frank Delea
ISBN 0-471-44652-1

(59km).[2] Motors were introduced and the use of incandescent lamps continued to increase. By 1886, the dc systems were experiencing limitations because they could deliver energy only a short distance from their stations since their voltage could not be increased or decreased as necessary. In 1885 a commercially practical transformer was developed that allowed the development of an ac system. A 4000 volt ac transmission line was installed between Oregon City and Portland, 13 miles away. A 112-mile, 12,000 volt three-phase line went into operation in 1891 in Germany. The first three-phase line in the United States (2300 volts and 7.5 miles) was installed in 1893 in California.[3] In 1897, a 44,000-volt transmission line was built in Utah. In 1903, a 60,000-volt transmission line was energized in Mexico.[4]

In this early ac period, frequency had not been standardized. In 1891 the desirability of a standard frequency was recognized and 60Hz (cycles per second) was proposed. For many years 25, 50, and 60Hz were standard frequencies in the United States. Much of the 25Hz was railway electrification and has been retired over the years. The City of Los Angeles Department of Water and Power and the Southern California Edison Company both operated at 50Hz, but converted to 60Hz at the time that Hoover Dam power became available, with conversion completed in 1949. The Salt River Project was originally a 25Hz system, but most of it was converted to 60Hz by the end of 1954 and the balance by the end of 1973.[5]

Over the first 90 years of its existence, until about 1970, the utility industry doubled about every ten years, a growth of about 7% per year. In the mid-1970s, due to increasing costs and serious national attention to energy conservation, the growth in the use of electricity dropped to almost zero. Today growth is forecasted at about 2% per year.

The growth in the utility industry has been related to technological improvements that have permitted larger generating units and larger transmission facilities to be built. In 1900 the largest turbine was rated at 1.5MW. By 1930 the maximum size unit was 208MW. This remained the largest size during the depression and war years. By 1958 a unit as large as 335 MW was installed, and two years later in 1960, a unit of 450MW was installed. In 1963 the maximum size unit was 650MW and in 1965, the first 1,000MW unit was under construction.

Improved manufacturing techniques, better engineering, and improved materials allowed for an increase in transmission voltages in the United States to accompany the increases in generator size. The highest voltage operating in 1900 was 60kV. In 1923 the first 220kV facilities were installed. The industry started the construction of facilities at 345kV in 1954, in 1964 500kV was introduced, and 765kV was put in operation in 1969. Larger generator stations

[2] *Ibid.*
[3] *Ibid.*
[4] *Ibid.*
[5] *Ibid.*

required higher transmission voltages; higher transmission voltages made possible larger generators.

These technological improvements increased transmission and generation capacity at decreasing unit costs, accelerating the high degree of use of electricity in the United States. At the same time, the concentration of more capacity in single generating units, plants, and transmission lines had considerably increased the total investment required for such large projects, even though the cost per unit of electricity had come down. Not all of the pioneering units at the next level of size and efficiency were successful. Sometimes modifications had to be made after they were placed in operation; units had to be derated because the technology was not adequate to provide reliable service at the level intended. Each of these steps involved a risk of considerable magnitude to the utility first to install a facility of a new type or a larger size or a higher transmission voltage. Creating the new technology required the investment of considerable capital that in some cases ended up being a penalty to the utility involved. To diversify these risks companies began to jointly own power plants and transmission lines so that each company would have a smaller share, and thus a smaller risk, in any one project. The sizes of generators and transmission voltages evolved together as shown in Figure 1.1.[6]

The need for improved technology continues. New materials are being sought in order that new facilities are more reliable and less costly. New technologies are required in order to minimize land use, water use, and impact on the environment. The manufacturers of electrical equipment continue to expend considerable sums to improve the quality and cost of their products. Unfortunately, funding for such research by electric utilities through the Electric Power Research Institute continues to decline.

DEVELOPMENT OF THE NATIONAL ELECTRIC POWER GRID[7]

Electric power must be produced at the instant it is used. Needed supplies cannot be produced in advance and stored for future use. At an early date those providing electric power recognized that peak use for one system often occurred at a different time from peak use in other systems. They also recognized that equipment failures occurred at different times in various systems. Analyses showed significant economic benefits from interconnecting systems to provide mutual assistance. The investment required for generating capacity could be reduced. Reliability could be improved. This lead to the development of local, then regional and subsequently three transmission grids which covered the United States. In addition, differences in the costs of producing

[6] J.A. Casazza, 1993, *The Development of Electric Power Transmission—The Role Played by Technology, Institutions, and People*, IEEE Case Histories of Achievement in Science and Technology, Institute of Electrical and Electronic Engineers.

[7] *Ibid.*

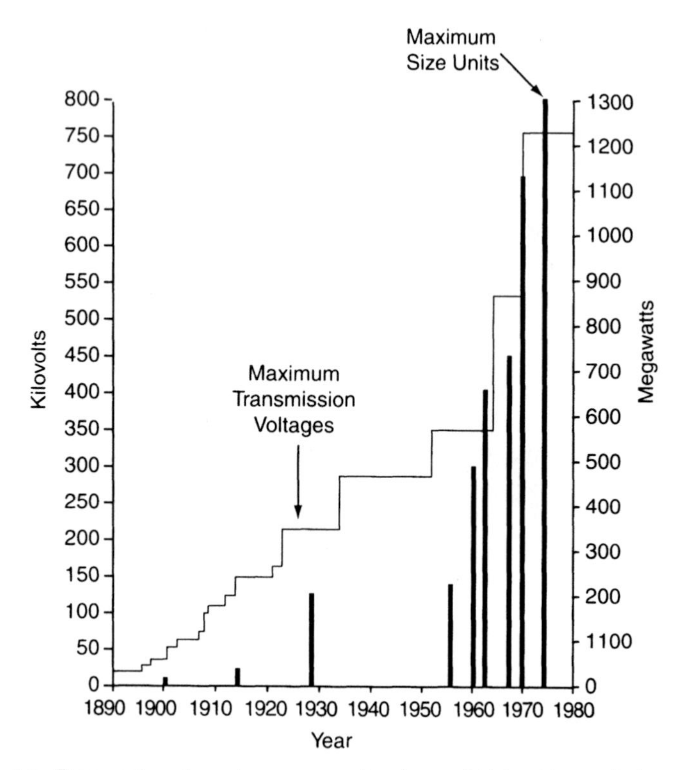

Figure 1.1. Progression of maximum generator size and highest transmission voltage.

electricity in the individual companies and regions often resulted in one company or geographic area producing some of the electric power sold by another company in another area. In such cases the savings from the delivery of this "economy energy" were usually split equally among the participants. Figure 1.2 shows the key stages of the evolution of this grid.[8] Figure 1.3 shows the five synchronous power supply areas currently existing in the United States and Canada.[9]

The development of these huge synchronous areas, in each of which all generation is connected directly and indirectly by a network of transmission lines (the grid), presents some unique problems because of the special nature of electric power systems. Whatever any generator or transmitter in the synchronous region does or does not do affects all others in the synchronous region, those close more significantly and those distant to a lesser degree. The loss of a large generator in Chicago can affect systems in Florida, Louisiana, and North Dakota. Decisions on transmission additions can affect other

[8] *Ibid.*
[9] *Ibid.*

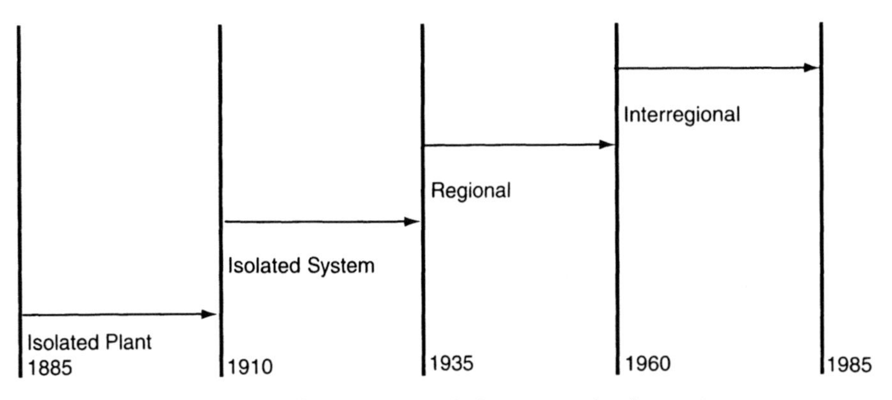

Figure 1.2. Stages of transmission system development.

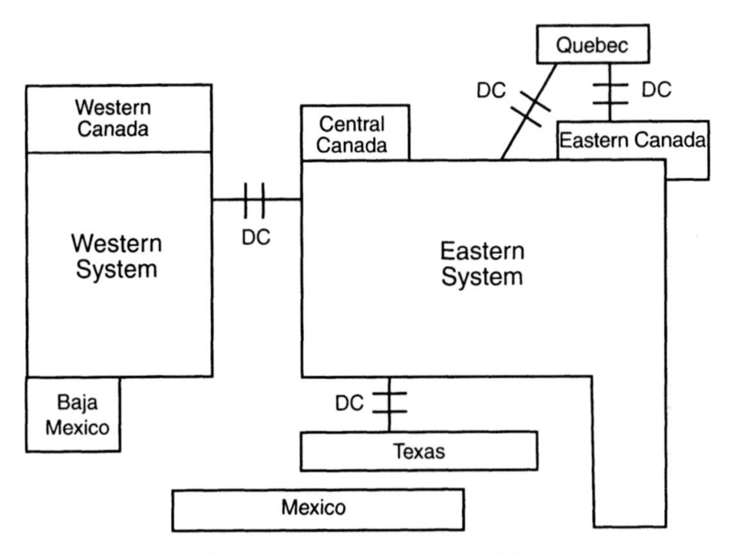

Figure 1.3. The five synchronous systems of North America.

systems many hundreds of miles away. This has required the extensive coordination in planning and operation between participants in the past. New procedures will be needed in the future.

As stated by Thomas P. Hughes of the University of Pennsylvania in the September, 1986 issue of CIGRE Electra:[10] "Modern systems are of many kinds. There are social systems, institutional systems, technical systems, and systems that combine components from these plus many more . . . An example of such a technological system . . . is an electric power system consisting not

[10] *Ibid.*

only of power plants, transmission lines, and various loads, but also utility corporations, government agencies, and other institutions . . . problems cannot be neatly categorized as financial, technical, or managerial; instead they constitute a seamless web . . . engineering or technical improvements also require financial assistance to fund these improvement(s) and managerial competence to implement them."

INDUSTRY OWNERSHIP STRUCTURE

At the turn of the century, the United States was dotted with approximately 5,000 isolated electric plants, each servicing a small area. Entrepreneurs bought these systems to form larger systems. It was easier to raise cash and savings could be obtained by coordinating generation, transmission, and the distribution system development over a wider region.

As shown a number of times in the electric power industry, because of its special nature, practices that lead to additional economics often lead to opportunities for additional abuses. The concentration of economic power in fewer and fewer organizations through highly leveraged purchases of companies led to Congress passing the Public Utility Holding Company Act of 1935.

Over more than 100 years the ownership of generation plants, transmission, and distribution systems has evolved. As shown in Figure 1.4, generation

Ownership Category	Number of Firms	Capacity gW	Percent of Total Industry Capacity
Investor Owned			
Integrated	140		
Generate & Transmit	10		
Transmit & Distribute	6		
Generate & Distribute	25		
Generate	11		
Transmit	7		
Distribute	34		
Other	6		
Total	239	415.7	51.2%
Rural Electric Co-ops	912	31.8	3.9%
Public, Non-Federal			
Municipal Utilities			
Public Utility Districts			
Irrigation Districts			
State and Mutual			
Total	2009	85.7	10.6%
Federally Owned	10	67.6	8.3%
Other		1.6	
Total Utility		602.4	74.2%
Non-Utility Generators		209.2	25.8%
INDUSTRY TOTAL		811.6*	

Figure 1.4. Ownership of the U.S. electric industry, utility and non-utility, 2000.

* EIA data for utility generation in 2000 indicates two values; 604,513 mW used in Figure 1.4 and 602,377 mW.

Fuel Type	Number of Units	Net Summer Capacity mW	Average Unit Size mW
Coal	1,024	260,990	255
Petroleum	3,007	41,017	14
Gas	2,068	117,845	57
Water – pumped storage	135	18,020	133
Water– conventional hydroelectric	2,836	73,738	26
Nuclear	91	86,163	947
Waste Heat	61	5,892	97
Other Renewable	128	837	7
Total Utility		604,502	
Total Non Utility		209,248	

Installed at Year-End 2000†

Fuel Type	Number of Units	Net Summer Capacity mW	Average Unit Size mW
Coal			
Petroleum	95	879	9
Gas	280	40,652	145
Water – pumped storage			
Water– conventional hydroelectric	24	373	16
Nuclear			
Waste Heat	13	1,425	110
Other Renewable			
Total Utility	458	44,726	

Planned Additions 2001-2005

Figure 1.5. U.S. electric utility generating capacity.‡

† Plants sold or transferred to nonutilities are not included in these data.

‡ Source EIA.

is currently owned by investor-owned companies, rural electric cooperatives, various non-federal governments, such as municipals, states, irrigation districts, and so forth, the federal government, and non-utility companies. Transmission systems are similarly owned. Toward the end of the 20th century many new participants became involved in the electric power industry, including merchant plant owners and power marketers. The capacity and energy sources for generators currently in service are shown in Figure 1.5.

As DeTouqueville observed centuries ago, the American genius is the ability to invent new organizations to meet our needs. The electric power industry formed trade organizations, for example, EEI, APPA, ELCON, and so forth, to lobby for various special interests; industry organizations such as NERC and reliability councils to insure reliability; EPRI and NYSERDA to do research; and professional organizations such as the IEEE, CIGRE, NARUC, and so on, to facilitate exchange of experience and new ideas among the professionals involved. (These are discussed in Chapter 16.)

LEGISLATION AND REGULATION

The golden age of electric utilities was the period from 1945 to 1965.[11] During this period there was exponential load growth accompanied by continual cost reductions. New and larger plants were being installed at a continuously lower cost per kilowatt reflecting economics of scale. Improvements in efficiency were being obtained through higher temperatures and pressures for the steam cycle, which was lowering the amount of fuel required to produce a kilowatt hour of electric energy. New generating plants were being located at the mine mouth, where coal was cheap, and power was transmitted to the load centers. This required new, higher-voltage transmission lines since it had been found that "coal by wire" was cheaper than the existing railroad rates.

The coordination of utilities was extensive. The leaders of the industry involved in planning the power systems saw the great advantage of interconnecting utilities to reduce capital investments and fuel costs. Regional and inter-regional planning organizations were established. The utilities began to see the advantage of sharing risk by having jointly owned units.

On the analytical side, improved tools were rapidly being developed. Greatly improved tools for technical analysis—such as computers—began to appear, first as analog computers and then as digital computers. At the same time, the first corporate models were developed for analyzing future plans for possible business arrangements for joint projects, of costs to the customers, for the need for additional financing, and the impact on future rates.

All of these steps reduced capital and fuel costs which resulted in lower rates. Everyone was happy. The customers were happy because the price of electricity was going down. The investors were happy because their returns on investments and the value of their stock were increasing. The system engineers were happy because they were working on interesting and challenging problems that were producing recognized benefits, and their value to the utility organizations was increasing. Finally, the business mangers were happy that they were running organizations that were functioning smoothly and were selling their product to satisfied customers.

Blackouts and the Reliability Crisis

The first blow to this "golden age" was the blackout of New York City and most of the Northeast, in 1965, which was caused by events taking place hundreds of miles away. The government reaction was immediate. Joseph C. Swidler was then Chairman of the Federal Power Commission. On orders from President Johnson, he set up investigative teams to look into the prevention of future blackouts. As a result, they wrote an excellent report called "Pre-

[11] *Ibid.*

vention of Power Failures" which is a classic to this day.[12] This report and a number of subsequent blackouts led to increasing attention by Congress, the FERC,[13] and the DOE to questions of reliability and increasing study. As an alternate to additional legislation, the industry recognized the need to govern itself and formed NERC and EPRI. Formal regional reliability criteria were developed, reliability conditions monitored and major funds contributed to develop new technology.

Environmental Crisis—The Shift to Low-Sulfur Oil

Starting shortly after the reliability crisis, and overlapping it considerably, was the environmental crisis. Both the public and the government became concerned about air quality, water quality, and the effect of electricity production on the environment. New environmental legislation was passed. These laws made the siting of new power plants very difficult. The power industry began installing nuclear units (which essentially had no exhaust); converting some of the existing coal-burning units to low-sulfur oil; providing electrostatic precipitators to filter-out particulate emissions; installing scrubbers to remove sulfur combustion products; and installing cooling towers so rivers would not heat up. All of these steps helped meet new government environmental requirements but significantly increased capital costs and fuel costs.

Fuel Crisis—The Shift from Oil

While these changes and additions were still underway, the industry was overtaken by another crisis. In 1973 the OPEC organization stopped all delivery of oil to the United States. This raised serious questions about plans to convert existing units to oil. Plans were cancelled to convert generation to oil (at a considerable financial penalty). Huge increases in the price of fuel occurred.

Financial Crisis

At the same time, the country found itself in an inflationary spiral; the annual cost of money rose to double digits rates. All utility costs increased rapidly, requiring large rate increases. Because of the political impacts of such rate increases, many state regulatory commissions rejected requests from the utilities for needed rate increases, thus exacerbating the financial problems of utilities. The depressed economy and rising costs of electricity dampened electric

[12] Federal Power Commission, Prevention of Power Failures, Volume I, Report to the Commission, Washington, D.C., July 1967.
[13] FERC—The Federal Energy Regulatory Commission—is the successor to the Federal Power Commission (FPC).

sales and load growth. The financial crisis resulted in a period of increasing costs, declining revenue, the lack of load growth, and large amounts of generating capacity under construction that would not be needed as soon as originally projected. Utilities were forced to cancel construction of projects already underway, resulting in large cancellation payments. In 1979 a major accident occurred at the Three Mile Island nuclear plant in Pennsylvania. Massive overruns occurred in the cost of nuclear plants still under construction as the Nuclear Regulatory Commission responded by requiring significant modification in designs. As a result, a large amount of planned nuclear generation was never built.

The service dates for other plants were delayed, in some cases for many years. This delay amplified the financial crisis even further because there was an appreciable investment in these partially completed plants on which earnings were required, even though the plants were not operating and producing any electricity. Tenfold cost increases were experienced by many of these plants.

Legislative and Regulatory Crisis

At about the same time, the Federal Government had become very chaotic and unpredictable in the regulations it issued. Some believed that paying to reduce peak power consumption was more economical than building new generating and transmission capacity. This concept has been called "least-cost," "demand-side," or "integrated resource" planning.

The Public Utilities Regulatory Power Act (PURPA) legislation passed in 1978, prescribed the use of "avoided costs" for determining payments to independently owned co-generators and qualifying facilities (QFs), such as low-head hydro and garbage burners. These "avoided costs" were the alternate utility costs for producing electricity based on the alternates available to the utility system. They were based on estimates of future costs, made by state regulators, which turned out to be much higher than the actual costs that occurred primarily because of the significant over-estimates of the future price of fuel. Unfortunately many utilities were required to sign long-term contracts for purchased energy reflecting these cost estimates. The avoided-cost approach led to excessive payments to some co-generators and other qualifying facilities. Subsequently, some utilities had to make very large payments to the plant owners to cancel such contracts or to purchase the plants.

The next step by the various regulatory commissions was the proposal, and in some cases the adoption, of competitive bidding procedures for new generators. One of these procedures called for competitive bidding for the provision of the electricity needed each hour. It required all bidders whose proposals were accepted to be paid the highest bid accepted for the hour, even though their proposal was lower. This approach caused huge additional costs. In many cases it is being replaced by negotiated contracts to buy specific amounts of power.

The 1992 Energy Policy Act, FERC Orders 888 and 889, and various other FERC orders and notices followed. Rapidly rising costs, declining reliability, developing procedures for manipulating electricity prices, have all increased concern and scrutiny of the electric power industry.

2

ELECTRIC POWER SYSTEM

This chapter gives an overview of the electric power system. The electric power industry delivers electric energy to its customers which they, in turn, use for a variety of purposes. While power and energy are related[1], customers usually pay for the energy they receive and not for the power.

An electric power system is comprised of the following parts:

- Customers[2], who require the electric energy and the devices in which they use the electric energy—appliances, lights, motors, computers, industrial processes, and so on;
- Sources of the electric energy—electric power plants/electric generation of various types and sizes;
- Delivery system, by which the electric energy is moved from the generators to the customers.

Taken together, all of the parts that are electrically connected or intertied operate in an electric balance. The technical term used to describe the balance is that the generators operate in synchronism with one another. Later we will

[1] See Chapter 3 for explanation of power and energy.

[2] Some have questioned inclusion of customers as a part of the power system. The authors feel that the magnitude, location, and electrical characteristics of customer load are as important as those of generators. Additionally, demand side management and distributed generation also impact both the electrical and commercial operation.

Understanding Electric Power Systems: An Overview of the Technology and the Marketplace, by Jack Casazza and Frank Delea
ISBN 0-471-44652-1

discuss how this concept of being in synchronism applies in the United States and Canada.

CUSTOMERS

Customer usage is typically referred to as customer load or "the load". The peak usage, usually measured over an hour, a half-hour, or 15 minutes (peak demand) is measured in either kilowatts or megawatts. The energy used by a typical residential or small commercial customer is measured in kilowatt-hours and that used by larger customers in megawatt-hours.

Industry practice has been to group customers by common usage patterns. Typically these customer groups (or classes) are:

- Residential;
- Commercial;
- Industrial;
- Governmental;
- Traction/railroad.

A reason for delineating customer types is to recognize the costs that each customer class causes in the provision of electric service since different customer classes have different usage patterns with differing impacts on the capital and operating costs. In a regulated environment, where customers are charged for their usage of electricity based on the cost of that supply, these classifications allow different menus of charges (rates) to be developed for each customer class.

Analyzing different customer types also facilitates forecasting changes in customer electric requirements. These forecasts are required for long-range planning and short-range operating purposes.[3]

Individual customer requirements vary by customer type and by hour during the day, by day during the week and by season. For example, a residential customer's peak hour electricity consumption will normally occur in the evening during a hot summer day when the customer is using both air conditioning, lighting and perhaps a TV, computer or other appliances. A commercial customer's peak hour consumption might also occur during the same day but during afternoon hours, when workers are in their offices.

The time of day when a system, company or geographic area peak occurs depends on the residential, commercial and industrial customer mix in the area. The aggregate customer annual peak demand usually occurs during a hot

[3] See Chapter 4, Electric Energy Consumption.

summer day or a cold winter day, depending on the geographic location of the region and the degree of customer use of either air conditioning or electric heating. The electric system is built to meet the maximum aggregate system and local area peak customer demand for each season.

Diversity refers to differences in the time when peak load occurs. For example, if one company's area is heavily commercial and another's is heavily residential, their peaks may occur at different times during the day or even in different seasons. This timing difference gives the supplying company the ability to achieve savings by reducing the total amount of capacity required.

The types of electric devices customers use also have an important bearing on the performance of the electric system during times of normal operation and times when electrical disturbances occur such as lightning strikes, the malfunctioning and loss of generating resources or damage to parts of the delivery system. Some types of customer equipment can require that devices be installed to provide extra support to maintain the power system's voltage.

The electric system has metering equipment to measure and record individual customer electric usage (except for street lighting) and systems to bill and collect appropriate revenues. For most customers, the meters measure an aggregate energy usage. For larger customers (usually commercial and industrial), meters also are used which record peak demand.

SOURCES OF THE ELECTRIC ENERGY—GENERATION

There are a number of ways to produce electricity, the most common commercial way being the use of a synchronous generator driven by a rotating turbine. The combination is called a turbine-generator.

The most common types of turbine generators are those where a fossil fuel is burned in a boiler to produce heat to convert water to steam which drives a turbine. The turbine is attached (coupled) to the rotating shaft (armature or rotor) of a synchronous generator where the rotational energy is transformed to electrical energy. In addition to the use of fossil fuels to produce the heat required to change the water to steam, there are turbine generators which rely on the fission of nuclear fuel to produce the heat. Other types of synchronous generators are those where the turbines are driven by moving water (hydro turbines) and gas turbines which are turned by the exhaust of a fuel burned in a chamber containing compressed air.

For each type system, there are many variations incorporated in the power plant in order to improve the efficiency of the process. Hybrid systems are also in use; an example is a combined cycle system where the exhaust heat from a gas turbine is used to help provide heat for a steam driven turbine. Typically, more than one these generating facilities were built at the same site to take advantage of common infrastructure facilities, that is, fuel-delivery systems, water sources and convenient points to connect to the delivery system.

Energy Source	Utilities Million - mWhrs	Non Utilities Million mWhrs	Total mWhrs	% of GRAND TOTAL
Coal	1,696.6	271.1	1,967.70	51.8%
Petroleum	72.2	36.6	108.80	2.9%
Gas	290.7	321.7	612.40	16.1%
Nuclear	705.4	48.5	753.90	19.8%
Hydroelectric	248.2	24.9	273.10	7.2%
Geothermal	0.2	14.0	14.20	0.4%
Other	2.1	67.8	69.90	1.8%
GRAND TOTAL	3,015.4	784.6§	3,800.00	100.0%

Figure 2.1. Energy sources of utility and non-utility generation in 2000.**
§ Non Utility Value is preliminary.
** Source EIA Table 3 and 58, U.S. Electric Utility and Non Utility Generation 1990 through June 2002.

A small, but not insignificant, segment of the electric generation in the country includes technologies that are considered more environmentally benign than traditional sources; that is, geothermal, wind, solar, biomass. In many of these technologies dc power is produced and use is made of inverters to change the dc to the alternating current (ac) needed for transmission and use.

Figure 1.4 shows a total of 811,625 megawatts of utility and non-utility electric generating capacity in the United States in 2000. Figure 2.1 shows the various energy sources used.

Generators are selected, sized and built to supply different parts of the daily customer load cycle. One type generator might be designed to operate continuously at a fixed level for the entire day. This is a base loaded generator. Another generator might be designed to run for a short period at times of peak customer demand. This is a peaking generator. Others might be designed for intermittent type service.

One important aspect of the selection of a particular generator is the trade off between its installed cost and its operating costs. Base loaded generators have much higher installed costs per unit of capacity than peaking generators but much better efficiency and lower operating costs. Included in this decision is the availability and projected cost of fuel.

Prior utility practice has been to have enough generation available to meet the forecast customer seasonal peak demand plus an adequate reserve margin. Reserve margins were determined by conducting probability studies considering, among other things, the reliability of the existing generation and potential future loads. Systems that were mainly hydro generation based had lower reserve margins (~12%) than systems that had nuclear, coal, or oil fired generation (~16–24%). The availability of aid from neighboring systems during shortages also had a large impact on the required reserve.

DELIVERY SYSTEM

A system of overhead wires, underground cables and submarine cables is used to deliver the electric energy from the generation sources to the customers. This delivery system, which electrically operates as a three phase, alternating current system, has four parts:

1. Transmission;
2. Subtransmission;
3. Primary distribution;
4. Secondary distribution.

The wires that make up the three phases are collectively called a line, circuit, or with distribution, a feeder.

The characteristic which differentiates the four parts of the delivery system from one another is the voltage at which they operate. In any one region of the country, transmission operates at the highest voltages, subtransmission at a lower voltage, then the primary distribution followed by the secondary distribution.

There is no uniformly agreed upon definition of what voltages constitute the transmission system. Some organizations consider voltage levels of 230 kV and above while others consider voltage levels of 115 kV and above. Figure 2.2 shows the voltages that generally are considered for each grouping in the United States.

The transmission systems in the various parts of the United States have different characteristics because of differences in the locations of generating units and stations in relation to the load centers, differences in the sizes and types of generating units, differences in geography and environmental conditions, and differences in the time that the transmission systems were built. Due

System	Voltages Included
Transmission††	765kV, 500kV, 345kV 230kV, 169kV, 138kV, 115kV
Subtransmission	138kV, 115kV, 69kV, 33 kV, 27 kV
Primary distribution	33kV, 27kV, 13.8kV, 4kV
Secondary distribution	120/240 volts, 120/208 volts, 240/480 volts

Figure 2.2. Classification of voltages in the United States.
†† In addition to the listed voltages, there are a number of high-voltage direct current (HVDC) installations that are classified as transmission.

Voltage	Miles of Transmission Line
ac	
230 kV	76.762
345 kV	49,250
500kV	26,038
765 kV	2,453
Total AC	154,503
dc	
250-300 kV	930
400 kV	852
450 kV	192
500 kV	1,333
Total DC	3,307
Total ac & dc	**157,810**

Figure 2.3. 1999 transmission circuit miles.‡‡
‡‡ Source—NERC.

to these differences, we find different transmission voltages in various sections of the country, that is, in some areas there is 765 kV, in others 500 kV and in others 345 kV.

As the industry developed, generation sites were usually located away from high-density customer load centers and the high-voltage transmission system was the most economic and reliable way to move the electricity over long distances. When new large central station generating plants were built, they either were connected to the nearest point on the existing transmission system or they were the trigger to institute the construction of transmission lines at a new higher transmission voltage.[4] The connection points are called substations or switching stations. These new higher voltage lines were connected to the existing system by means of transformers. This process is sometimes referred to as an overlay and resulted in older generation being connected to transmission at one voltage level and newer, larger generation connected at a new higher voltage level. Over time, the lower voltage facilities became called subtransmission.

The progression of transmission voltage levels in the United States in the 20th century is shown in Figure 1.1. As shown in Figure 2.3, in 1999 there were almost 154,500 miles of HVAC transmission operating at a voltage of 230 kV or higher in the United States.

Transformers enable the wires and cables of different voltages to operate as a single system. A transformer is used to connect two (or more) voltage levels.[5]

[4] Remember that the electric system involves a large capital investment for facilities with service lives measured in decades. Changes to the system are incremental to that which already exists.
[5] Transformers are explained in Chapter 3.

Transformers are installed at the generating plant to allow the generators, whose terminal voltage is typically between 13kV and 24kV, to be connected to transmission. These are called generator step-up transformers. As the delivery system brings the electricity closer to the customers, transformers connect the higher voltage system to lower voltage facilities. Connections can be made to the local subtransmission system or directly to the primary distribution system. These are step-down transformers. Figure 2.4 shows a conceptual sketch of a power system.

The connection point between the transmission system or the subtransmission system and the primary distribution system is called a distribution substation.[6] Depending on the size of the load supplied, there can be one or more transmission or subtransmission lines supplying the distribution substation. A distribution substation supplies a number of primary distribution feeders. These distribution feeders can supply larger customers directly or they connect to a secondary distribution system through a transformer affixed to the top of a local utility pole or in a small underground installation.

Depending on the magnitude of their peak demand, customers can be connected to any of the four systems. Typically a residential customer will be connected to the secondary distribution system. A commercial customer, that is, a supermarket or a commercial office building, will normally be connected to the primary distribution system. Very large customers such as steel mills or aluminum plants can be connected to either the subtransmission or transmission system.

INTERCONNECTIONS

As individual companies built their own transmission, it became apparent that there were many reasons to built transmission lines or interties between adjacent systems. Among the reasons were:

- Sharing of generation reserves thereby reducing the overall amount of generating capacity and capital investment needed;
- Providing the ability to buy and sell electricity to take advantage of differences in production costs;
- Facilitating operations by allowing more optimum maintenance scheduling;
- Providing the ability to jointly construct and own power plants;
- Providing local load support at or near the company boundaries.

[6] The substations are sometimes called switching stations. In addition, substations are also called high voltage substations, bulk power substations, and distribution substations.

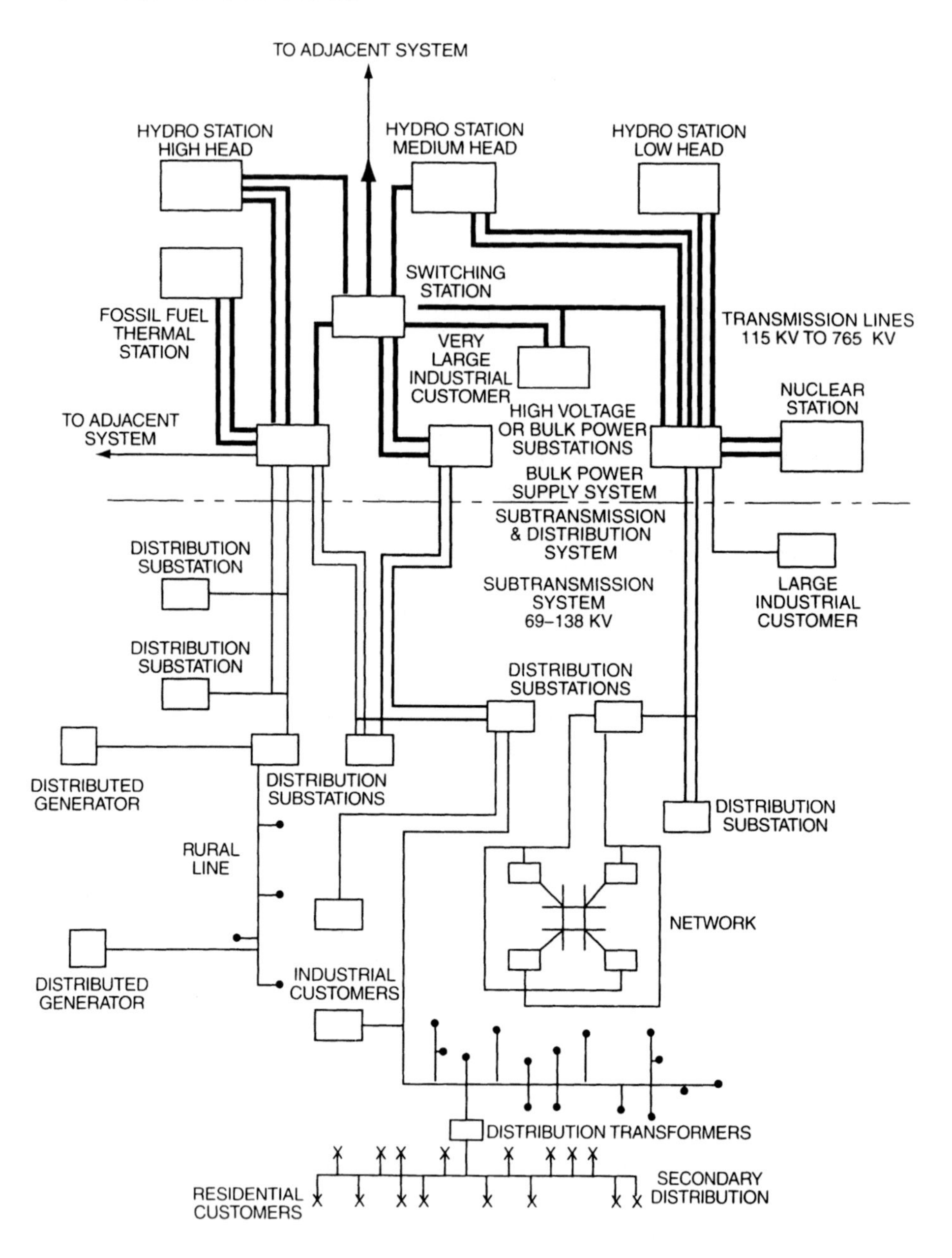

Figure 2.4. Conceptual sketch of an electric system.[§§]

[§§] From *Electric Utility Systems and Practices*, *4th Edition*, Homer M. Rustebakke, Wiley, New York.

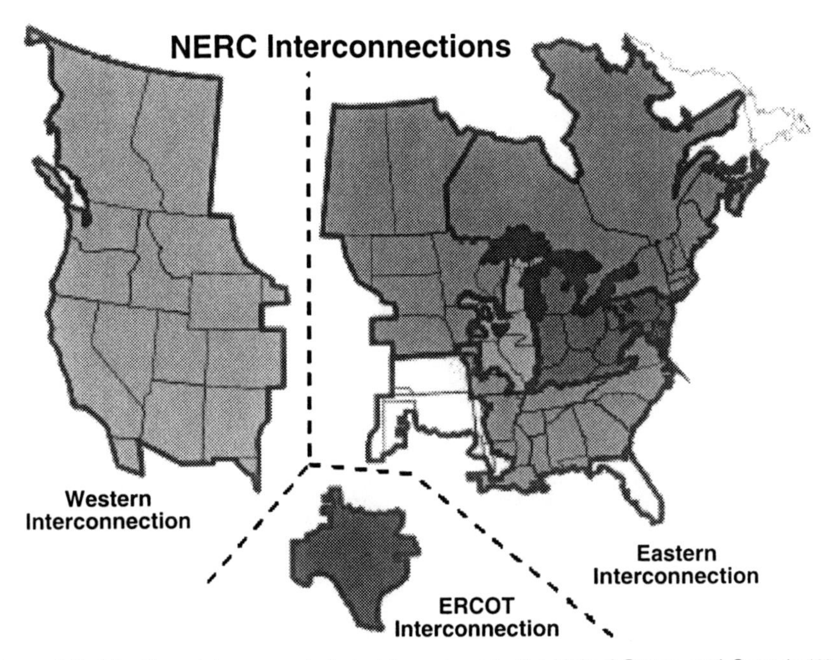

Figure 2.5. The three interconnected electric systems in the United States and Canada.***
*** Source—NERC.

GRID

The resulting transmission system is not a linear arrangement of lines fed from a single generating station and tying to a single primary distribution system, but something much more complex. Generating units are located at a number of sites as are the locations of distribution substations. The generating sites are often electrically directly connected by transmission lines (some short and some long) to nearby substations where transmission lines also connect. Other transmission lines connect the substations together and also connect to distribution substations where there are connections to lower voltage facilities. From some of the substations there are interconnections to other companies.

Taken together, this arrangement of transmission lines tied together at various substations provides a degree of redundancy in the delivery paths for the electric energy. Power engineers have coined the terms "the grid", the "bulk power system" and "the interconnection" to describe the delivery system. Figure 2.5 shows the three large grids in the United States and Canada: the Eastern Interconnection, the Western Interconnection and ERCOT, the Texas system. The generators within each grid operate in synchronism with one another. The Canadian Province of Quebec is interconnected to the eastern United States grid by non-synchronous HVDC ties.

3

BASIC ELECTRIC POWER CONCEPTS

This chapter describes, as simply as possible, the applicable physical laws and concepts needed to understand the physical operation of an electric power system. In Chapters 10 to 17 the commercial operation of the system is covered. An attempt has been made to present the material in a nontechnical (i.e., as few equations as possible) manner.

It is important to remember that the operation of an electric power system is governed and described by the laws of physics, which are unchanging, whereas the commercial operations are subject to man-made rules which are subject to modification and change. There is an interrelationship between the two in that the rules established for commercial operations must recognize and respect the physical laws by which the power system operates and the commercial rules often determine the design and operation of the system.

A note on terminology is warranted. As the electric utility industry developed, an associated jargon evolved, some of which you have already been exposed to in Chapter 2. In some instances, the words used are simply contractions of longer terms; for example, the use of "amps" in place of "amperes" to describe electrical current. In other instances, a variety of terms are used to describe a single concept, for example, the use of the terms "grid", "the interconnection", and "the bulk power system" to describe the totality of all the electric transmission.[1]

[1] There are many glossaries defining the terms used in connection with electric power systems and business. One of the most useful is *CIGRE Glossary of Terms Used in the Electricity Supply*

Understanding Electric Power Systems: An Overview of the Technology and the Marketplace, by Jack Casazza and Frank Delea
ISBN 0-471-44652-1

ELECTRIC ENERGY

Energy is defined as "the capacity for doing work." Electricity is but one of many forms of energy. Other familiar forms of or descriptions of energy are thermal or heat, light, mechanical, and so on. Energy is also described as kinetic, that energy associated with a moving body, and potential energy, that energy associated with an object's position.

For centuries mankind has used energy in its various forms to enhance its standard of living. In many cases, ways have been devised to change energy from one form to another to increase its usefulness. An example as old as mankind is the burning of a fuel to produce heat and light. Electric power systems thus provide a service, energy in a usable form, not a product, to consumers.

Electrical energy possesses unique characteristics that made it an extremely valuable form of energy. It has unique properties:

- It can be produced at one location and transmitted to another instantaneously;
- It can be transformed to other energy forms and thereby used in a variety of ways;
- It can be delivered by a system of wires, and control;
- It cannot be stored.

Consistent terminology has always been an issue when discussing electricity and electric power. Convention is to use a system of measurement based on the MKS (meter, kilogram, second) system.

Figure 3.1 summarizes the terms used to describe various aspects of electricity and shows some of the interrelationships between them. There is an electric charge associated with electrons. This charge is described by a quantity called a coulomb. The rate of flow of these charges is called the electric current and is described by a quantity called an ampere. One ampere is equal to the flow of one coulomb of charge during one second across a reference point. The capital letter, I, is used to indicate current and the quantity is sometimes referred to as amps. In many texts, electric current is described as a physical flow of electrons. It is not. The electrons do not flow.[2] Rather electricity is a flow of energy as a result of electron vibrations. The mechanism is the transfer of energy from one electron to another as they collide, one with another.

Electromagnetic force (EMF), voltage, and difference in potential are different descriptions of the notion of what causes these charges to flow. A physics text would define voltage as the energy per a unit of charge where

Industry, February 2002. This includes the definitions used by the IEEE, NERC, EEI, and in Europe.

[2] "Amicus Brief to Supreme Court", August 2001, *www.tca-us.com*.

Quantity	Name or Unit	Symbol	Relationships
Electric charge	Coulomb	q	
Time	Seconds, Hours	t	
Current	Amperes	I	$I = q / t = V / R$
Resistance	Ohms	R	$R = V / I$
Inductive Reactance	Ohms	X_L	$X_L = 2 * \pi * f * L$
Capacitive Reactance	Ohms	X_C	$X_C = 1 / (2 * \pi * f * C)$
Impedance	Ohms	Z	$Z = R + j (X_L + X_C)$
Voltage	Electromagnetic force (EMF), Volts, kilovolts	E, V, kV	$V = I * R$ $V = J / Q$
Power or Real Power	Watts, kiloWatts, megaWatts,	P	$P = V * I$ $P = I^2 * R$ $P = V^2 / R$
Reactive Power	VArs, kiloVArs, megaVArs	Q	$Q = I^2 * X_L$ $Q = I^2 * X_C$
Apparent Power		S	$S = P + j Q$
Energy	kiloWatthours, megaWatthours, Joules	J	$J = V * I * t$ $J = I^2 * R * t$
Frequency	Hertz, cycles per second	f	

Figure 3.1. Basic electric relationships.

energy is measured by a quantity called a watt-second.[3] An engineering text would say that a difference in potential (or of voltage) of one volt causes a current of one ampere to flow through a circuit that has a resistance of one ohm.

The letter E is used when referring to a voltage source such as a generator or a battery and is often called an electromagnetic force. The letter V is used in all other instances. In both cases the quantity is measured by a quantity called a volt.[4] One volt is equal to one watt-second of energy per one coulomb of charge.

Voltage can be thought of as electric potential to deliver energy.[5] Differences in voltage measure the work that would have to be done[6] to move a unit charge from a point of one voltage to that of another voltage.

[3] In physics texts, the term joule is also used for energy. One joule = 1 watt-second.

[4] Named after Alessandro Volta, (1745–1827), an Italian who invented the battery.

[5] The voltage associated with an electric generator is called an electromagnetic force or EMF.

[6] This work would be done against the electric field.

When a source of voltage is applied to a wire, a current will flow. The material in the wire offers some resistance to the flow of current. This resistance is described by a quantity called an ohm.[7] One ohm is defined as the resistance of a circuit element when an applied voltage of one volt results in a current of one ampere. The resistance of wire depends on the material it is made of, the cross-sectional area of the wire and its length. For a given material, the larger the cross-sectional area the lower the resistance. The letter R is used to represent resistance in ohms. The relationship between voltage, current and resistance is known as Ohm's Law; that is, voltage = current × resistance. This relationship is applicable for direct current conditions. Later in this chapter we will introduce a version of this law when it is applied to alternating current circuits of the type used in power systems.

As mentioned earlier, electricity is a form of energy which is measured by a quantity called a watt-second. Electric customers usually see their bills keyed to their watt-hour usage. A related but different quantity is electric power. The unit of power is the Watt.

In an electric power system the magnitudes of many quantities are such that larger units are needed to describe them. The larger increments usually encountered are described by the addition of the term kilo or the term mega to the base unit. For example, a kiloVolt is 1,000 Volts; a kiloWatt is 1,000 Watts, a megaWatt is 1,000,000 Watts or 1,000 kiloWatts, a kiloWatthour is 1,000 Watthours.

CONCEPTS RELATING TO THE FLOW OF ELECTRICITY

The two laws which define the flow pattern of electricity through electric lines are known as Kirchhoff's Laws[8]. These laws reflect two basic physical concepts:

1. The algebraic sum of voltage drops around a closed loop is zero. The voltage across a source is considered positive while the voltage drop across an element (i.e. a resistor) is considered negative.
2. The algebraic sum of currents entering any common point (a node) where three or more lines connect must equal the algebraic sum of the currents leaving that point.

In a complex network of lines such as one finds in a power system, the flow of power and current in any one line is determined by the line's electrical characteristics, the characteristics of the other lines in the network, the location of the power injections and the locations of the power deliveries. All lines operating in a network participate to a greater or lesser extent whenever there

[7] Named after George Simon Ohm, (1787–1854), a German physicist.

[8] Named after Gustav Robert Kirchhoff, (1824–1887), a German physicist.

is an increment in generation to supply an increment of load. Later in this chapter we will discuss this notion in more detail.

DIRECT CURRENT

The first utility systems installed by Edison used direct current technology. The electrical energy in a direct current system is the same as found when a battery is used. If one looked at a picture of the voltage and the current, one would see that both had a constant, non-varying value. Not long after Edison installed his direct current system, others realized that the use of an alternating current system had advantages over the direct current. The concepts discussed hereto fore apply to direct current systems.

ALTERNATING CURRENT

The modern electric power system is an alternating current, three phase system. Electricity is generated by synchronous generators which are machines which convert the rotational energy of a shaft into electrical energy. The generator shaft or rotor is rotated by means of a turbine as discussed in Chapter 5.

The energy conversion is based on a phenomenon associated with magnetism and electricity called induction. If a stationary wire loop is placed in the field of a rotating magnet, an electric current will be induced in the wire. The rotor of an electric generator is made to look like a magnet by energizing conductors embedded in it with a source of direct current. The system that provides direct current to the rotor windings is called the excitation system. The energized windings on the rotor are conventionally called the field or field circuit. In modern generators the direct current excitation is derived from an alternating current source that has been rectified to provide dc.

The direct current excitation establishes a magnetic field in the metal of the rotor which extends across the air gap between the rotor and the stationary part of the generator (stator or armature). Electricity is induced in coils which are placed in slots in the stator. The voltage induced in any one coil reflects the time varying characteristic of the magnetic field, as viewed by a stationary observer, caused by the rotation of the rotor. The magnitude of the induced voltage can be adjusted up or down by changing the magnitude of the direct current flowing in the rotor. This is done by a voltage regulator in the excitation system which monitors the voltage at the terminal of the electric generator and adjusts the field voltage up or down as required to maintain the desired generator terminal voltage.

The voltage and current have a sinusoidal shape, that is, in each cycle of 360 degrees, it starts at a zero value at zero degrees, rises to a positive maximum at 90 degrees, falls to zero at 180 degrees, continues to fall to a negative

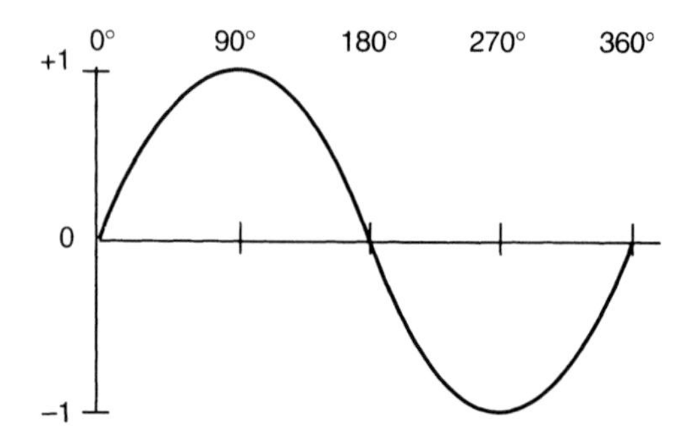

Figure 3.2. Sinusoidal shape of voltage or current.

maximum at 270 degrees and returns to zero at 360 degrees, where the process repeats as shown in Figure 3.2. This sinusoidal shape reflects the rotating pattern of the magnetic field produced on the rotor. If the stator coil is connected to an external load, current will flow. The current will also be oscillatory in nature, hence the name alternating current. The number of full cycles that occur in a set time defines the frequency of the electricity. In the United States and many other areas of the world, the frequency is 60 hertz[9] or cycles per second. In other areas a frequency of 50 cycles is used. The frequency is set by the number of magnetic circuits that are established on the rotor. The frequency of the electricity produced by a particular generator is defined as:

$$f = (n * p)/60^{10},$$

where n is the speed in revolutions per minute (rpm) and p is the number of pairs of magnetic poles. Steam turbines rotate at high speeds. For example, if one magnetic circuit is established, that is, there two magnetic poles established (a single pair consisting of a north and a south pole), a speed of 3,600 rpm will result in a frequency of 60 hertz. Alternately, if two magnetic circuits are established using two pairs of poles, a speed of only 1,800 rpm is needed to produce a frequency of 60 hertz. Hydraulic turbines rotate at relatively low speeds and will have many poles to produce the required frequency.

Because of the oscillatory nature of the voltage and current, an "effective" voltage[11] and current value is defined. These effective values are considered to be equivalent to the direct current voltage and currents that would produce the same power dissipation (as heat) in a resistance. The effective value for a sine wave is equal to 0.707 × the peak value. In the United States, for example, the oft quoted household voltage of 120 volts is an effective value and corresponds to a peak value of 169.7 volts.

[9] Named after Heinrich R. Hertz, (1857–1894), a German physicist.
[10] In this text, an asterisk (*) is used to indicate multiplication.
[11] The term root mean squared (RMS) is also used.

THREE PHASES

The efficiency of the energy transformation and delivery process improves as the number of independent coils located on stator is increased. The electrical conductors on the stator are physically arranged so that three separate but equal voltages are produced. The three conductors are connected together at a common point resulting in a configuration called a wye. The rotating patterns of these voltages are displaced from one another by 120 degrees. Taken together, the voltages and currents in the three coils become the three phases of a single circuit.

In a wye connection, two voltage measurements are defined: the phase voltage of the phase with respect to the neutral point, V_{LN}, and the voltage between phases or between lines, V_{LL}. When the system is balanced, that is, generators and customer load connected in such a way as to result in equal voltages and currents on each phase of a circuit, these two voltages are related as follows:

$$V_{LL} = \sqrt{3} * V_{LN} = 1.732 * V_{LN}$$

The convention used when referring to power system voltages is to use the effective or RMS value of the line to line voltage.

SYNCHRONISM

When a number of generators are connected to the same electric grid, they are said to be in synchronism because they operate at the same frequency and the angular differences between the voltage angles of each generator are stable and less than 90 degrees. Units operating in synchronism are magnetically coupled by their connections through the power system. If any one changes its angle of operation, all the others are affected.

CHARACTERISTICS OF AC SYSTEMS

Resistance

In an ac system, the voltage across a resistor and the current flowing thought it are said to be in phase, that is, their zero value and their maximum values occur at the same times. There are two types of fields associated with an ac electric system; electric fields and magnetic fields. Electric fields relate to the voltage and magnetic fields relate to the current.[12] The waveforms

[12] Since they operate at high voltage, transmission lines generate strong electric fields. If they are heavily loaded, i.e., carrying large amounts of power, they also will have strong magnetic fields. Distribution lines, which operate at lower voltages, generate weak electric fields, but can generate strong magnetic fields depending on the local customer load levels.

of the voltage and current associated with both of these characteristics are not in phase, that is, the times of the maximum and zero values are not identical.

Induction and Inductive Reactance

When we discussed the operation of a generator, we noted that an electric voltage is induced in a wire when a moving magnetic field "cuts" that wire. Similarly, a current varying with time (an alternating current) will produce a magnetic field around the wire carrying the current. Since the current is varying so will the magnetic field. This varying magnetic field "cuts" the conductor and a voltage is induced in the wire which acts to impede the originating current.

The relationship between the current and the induced voltage is defined by a quantity called the inductance. One Henry is the amount of inductance required to induce one volt when the current is changing at the rate of one ampere per second. The letter L is used to represent the inductance in Henries.

The inductance, L, of one phase of a transmission or distribution line is calculated by considering the self-inductance of the individual phase conductor and the mutual inductance between that phase and all other nearby phases both of the same circuit/feeder and other nearby circuits/feeders. These quantities are calculated based on the physical dimensions of the wires and the distances between them. The induced voltage across an inductor will be a maximum when the rate of change of current is greatest. Because of the sinusoidal shape of the current, this occurs when the actual current is zero. Thus the induced voltage reaches its maximum value a quarter-cycle before the current does—the voltage across an inductor is said to lead the current by 90 degrees, or conversely, the current lags the voltage by 90 degrees.

Inductive Reactance

The inductive reactance, X_L is a term defined to enable us to calculate the magnitude of the voltage drop across an inductor. The inductive reactance is measured in Ohms and it is equal to $2 * \pi * f * L$, where $2\pi f$ is the rotational speed in radians per second; π is called pi and its value is 3.1416, f = frequency in hertz and L = inductance in Henries. Inductances consume reactive power or VARs equal to I^2X_L.

Capacitance and Capacitive Reactance

An electric field around the conductor results from a potential difference between the conductor and ground. There is also a potential difference between each conductor in a three phase circuit and with any other nearby transmission lines. The relationship between the charge and the potential difference is defined by a quantity called the capacitance. One Farad is the amount of capacitance present when one coulomb produces a potential difference of one volt. The letter F is used to represent the capacitance in Farads.

The capacitance C, depends on the dimensions of the conductor and the spacing between the adjacent lines and ground. Since the charge on a capacitor varies directly with the voltage, when an alternating voltage is impressed across a capacitor, the flow of charge (or current) will be greatest when the rate of change of voltage is at a maximum. This occurs when the voltage wave crosses the zero point. Thus in an alternating current system, the current across a capacitor reaches its maximum value a quarter-cycle before the voltage does—the voltage is said to lag the current by 90 degrees, or conversely, the current leads the voltage by 90 degrees.

Capacitive Reactance

The capacitive reactance, X_C, is a term defined equal to $1/2 * \pi * f * C$, where C = capacitance in Farads. The unit of the capacitive reactance is Ohms. In a power system the capacitive reactance is viewed as a shunt connecting the conductor to ground. Capacitors supply reactive power or VARs equal to I^2Xc. Figure 3.3 demonstrates the current and voltage relationships for a resistor, an inductor and a capacitor.

Reactance

Both inductive reactance and capacitive reactance have an impact on the relationship between voltage and current in electric circuits. Although they are both measured in Ohms, they cannot be added to the resistance of the circuit since their impacts are quite different from that of resistance. In fact, their impacts differ one from the other. The current through an inductor leads the voltage by 90 degrees, while current through a capacitor lags the voltage by 90 degrees. Because of this difference, their effects will cancel one another. The convention is to consider the effect associated with the inductive reactance a positive value and that with the capacitive reactance a negative value and VARs as consumed by inductive reactance and supplied by capacitive reactance. A general term, reactance, is defined which represents the net effect of the capacitive reactance and inductive reactance. It is denoted by the capital letter X.

Impedance

Once determined, the reactance is combined with the resistance of a circuit to form a new quantity called Impedance which is denoted by the capital letter Z. To determine a single number representation of the impedance, the concept known as complex numbers is employed. Simply speaking, resistance and reactance are treated as both legs of a right triangle separated by 90 degrees. A common way of representing the impedance term is:

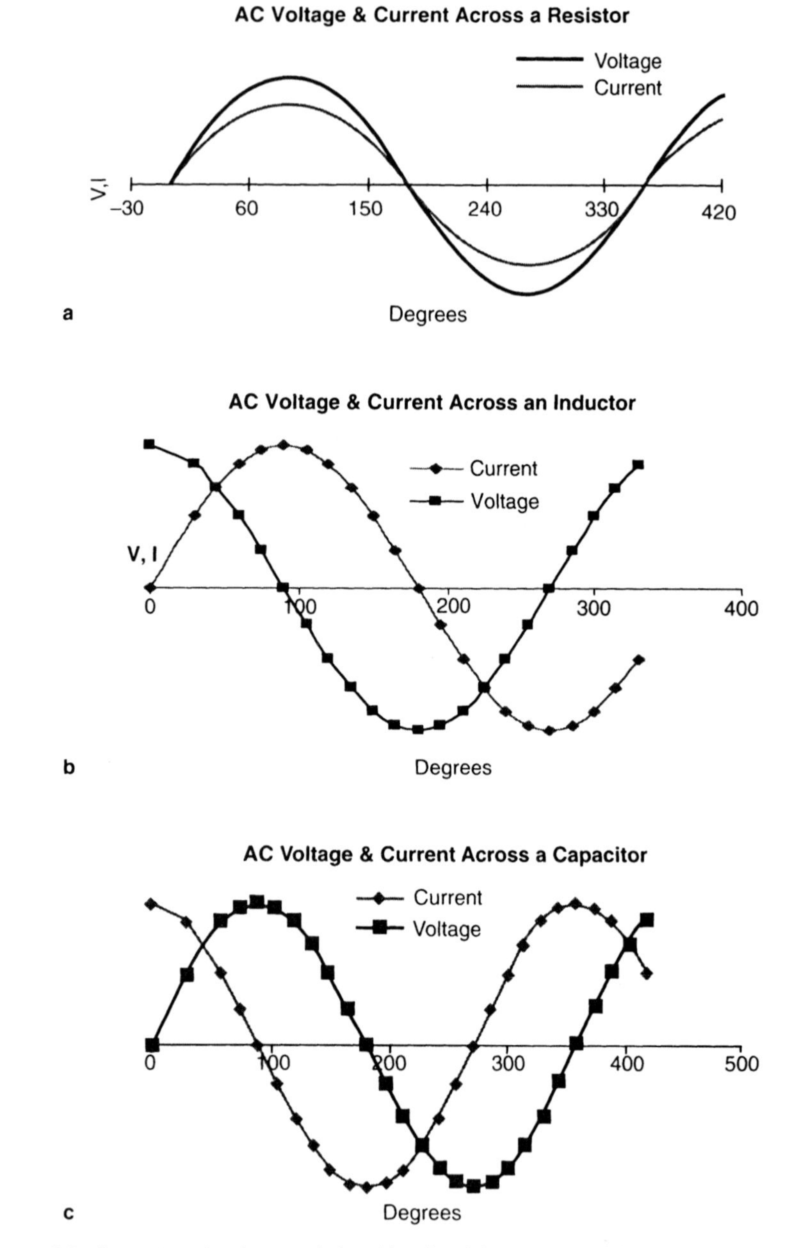

Figure 3.3. Current and voltage relationships for (a) a resistor, (b) an inductor and (c) a capacitor.

$$Z = R + j(X_L - X_C)$$

where the letter *j*, is used as a convention to indicate that the reactance is not to be directly added to the resistance. The magnitude of the impedance is determined by Pythagoras' Theorem, that is, the square of the impedance is equal to the sum of the squares of the resistance and the reactance.

$$Z^2 = R^2 + X^2, \text{or}$$

$$Z = (R^2 + X^2)^{0.5},$$

$$\text{where } X = X_L - X_C$$

Ohm's Law for Alternating Current

Ohm's Law as originally construed for dc cannot be applied to ac circuits since it recognizes only resistance and not the inductive and capacitive reactance effects. The Law can be modified to take into consideration the effect of reactance by simply replacing the term for the resistance with a term for the circuit's impedance and treating the voltage and the current as time varying quantities rather than as constants as in a direct current circuit. In engineering textbooks, the ac quantities are indicated by letters with lines drawn over them or by bold letters. We shall follow this latter convention.

$$\mathbf{V} = \mathbf{I} * \mathbf{Z}$$

POWER IN ALTERNATING CURRENT CIRCUITS

In a dc circuit, the power is equal to the voltage times the current, or P = V * I. This is also true in an ac circuit when the current and voltage are in phase; that is, when the circuit is resistive. But, if the ac circuit contains reactance, there is a power component associated with the magnetic and/or electric fields. The power associated with these fields is not consumed as it is in a resistance, but rather stored and then discharged as the alternating electric current/voltage goes through its cycle. This leads to another definition:

Apparent power = Real or True power (associated with a resistance)
+ Reactive power[13] (associated with an inductance or capacitance).

[13] Another name that has been used for this quality is Imaginary Power. The name derived from the application of the complex number convention to calculate Z.

A related concept is that of power factor, which is defined as a magnitude of P divided by a magnitude of S. In the electric power industry, if the power factor is too low (typically under 0.85) because of the magnitude of the reactive component Q, corrective actions are usually taken.

Using symbols:

$$S = P + jQ$$

Real Power

Real power is available to do work and is equal to the value of the resistance multiplied by the square of the current through the resistance. It is measured by a quantity called megawatts (mW) or kilowatts (kW).

$$P = I^2 * R$$

Reactive Power

Reactive power neither consumes nor supplies energy. The reactive power associated with an inductive reactance is the value of the inductive reactance multiplied by the square of the current through it. The reactive power is measured by a quantity called volt-ampere reactive or VARs. As the length of a line increases, its inductive reactance increases, and the more capacitive reactive power needed to offset the effect and to maintain adequate voltage:

$$Q_L = I^2 * X_L$$

The capacitive reactive power, Q_C, relates to the establishment of the electric field around a line. There are a number of ways to calculate this value, but the following offer insight into its effects on the transmission system.

$$Q_C = 3 * V^2{}_{LN}/X_C \text{ or } = \sqrt{3} * V_{LL} * I_C$$

In a power system, under normal operations, the voltage level on any one line is kept more or less constant, so the reactive power associated with the capacitance of the line is also relatively constant. Charging current, I_C, is defined as the line to neutral voltage divided by the capacitive reactance

$$I_C = V_{LN}/X_C$$

If the charging current becomes too large. much of the line's current carrying capacity may be "eaten up" by charging current. This situation sets limits on the length of an overhead line or of a cable that can be operated without installing some intermediate measures to offset the capacitive current. It is useful to visualize the impact of various devices on the reactive power of a power system as follows:

Sources of reactive power which raise voltage:

- Generators;
- Capacitors;
- Lightly loaded transmission lines due to the capacitive charging effect.

Sinks of reactive power which lower voltage:

- Inductors;
- Transformers;
- Most heavily loaded transmission lines due to the $I^2 * X_L$ effect;
- Most customer load (due to the presence of induction motors and the supply to other electric fields).

A synchronous generator can be made to be either a source of reactive power or a sink by using the generator excitation system to vary the level of its dc field voltage. During peak load conditions generators are usually operated to supply reactive power to the grid. During light load conditions generators may be used to absorb excess reactive power from the grid, especially where there are long transmission lines or cables nearby.[14]

ADVANTAGES OF AC OVER DC OPERATION

The use of ac allows the use of a multi-voltage level energy delivery system. High voltages are used for the transport of large blocks of power; lower voltages are used as smaller blocks of power are delivered to local areas; and the familiar 120/240 volt system is used for deliveries to individual customers. If the transmission of large amounts of electricity (or large blocks of power) were to take place using dc at the voltage levels normally found at the terminals of modern generators (13 kV to 30 kV), real power losses associated with the resistance of the transmission system would become prohibitive.

Use of dc for this purpose also would require that the supply voltage be the same, or close to the same, as that required by the equipment connected to the system. Considering the variety of types and sizes of electrical equipment; motors, lights, computers, and so forth, this is an impractical requirement.

[14] See generator capacity curves in Chapter 5.

TRANSFORMERS

The device which allows the use of a multi-voltage level energy delivery system is a transformer. A transformer makes use of the principal of electromagnetic induction that we've seen before. Figure 3.4 contains a simplified example showing two independent circuits wound about a common core. A magnetic path, B, is established in the core by the ac in one of the circuits.

It can be shown that because of the time varying nature of the magnetic field created by the current in coil 1, an ac voltage is induced in the second coil (coil 2). The induced voltage is related to the ac voltage impressed on the first coil (coil 1) multiplied by the ratio of the number of turns (N_2/N_1) that the respective circuits make around the core.

Symbolically:

$$V_2 = V_1 * (N_2/N_1)$$

Depending on the number of turns of wire constructed into the transformer, the voltage level on the secondary side can be increased or can be decreased from that on the primary side. This capability offers significant flexibility in the design of a power system.

In an ideal situation (no losses associated with the transformer), the power entering the transformer is equal to the power exiting it:

$$V_1 * I_1 = V_2 * I_2$$

A significant reduction of real power line losses can be achieved by using a transformer to raise the transmission voltage and thereby lowering the current resulting in lower $I^2 \cdot R$ losses.

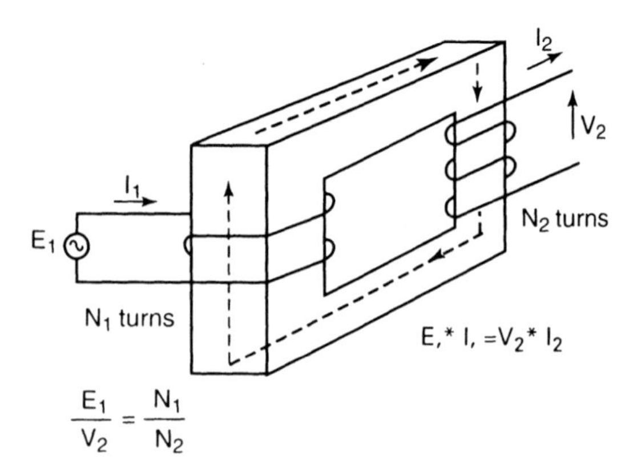

Figure 3.4. Conceptual schematic of a simple transformer.

POWER FLOW

By using Kirchhoff's Laws and the concept of impedance, calculations can be made to determine the currents, and real and reactive powers flowing in transmission lines and the voltages at all terminals (buses, nodes) in an ac system. Textbooks are written to educate students in the techniques of performing the required calculations. For our purposes we will summarize the insights that one gets from doing those analyses.[15]

Division of Power Flow Among Transmission Lines

Power flows on all possible transmission paths between a source of generation and a load approximately in proportion to the relative impedances of all the paths, not only on the most direct path.[16] It will not flow just on a path for which a commercial arrangement has been made (contracted path), nor only on the facilities of the owner or purchaser of the electricity.

For a vertically integrated utility, electricity from its generating plants will use all parallel paths to reach its load—including transmission facilities of neighboring utilities. When interregional power transfers are scheduled, the facilities of many utilities, not only those involved in the transaction, may carry some of the power. This phenomenon will always occur, no matter who owns the power plants or the transmission lines.

Voltage Drop and Reactive Power Flow

The magnitude of the voltage drop across a transmission line will depend predominately on the amount of reactive power (Q) that flows through the reactance of the line.[17] This means that to minimize a voltage drop, reactive power sources should be located as close as possible to the load. This relationship is summarized in the following formula:

$$(V_1 - V_2) = X_L * Q/V_2$$

Power Flow and Phase Angle Differences

The magnitude of power flow on a power system is dependent predominately on the phase angle difference between two points since the line reactance is constant and the terminal voltages V_1 and V_2 are essentially constant:

[15] In many instances the resistance of elements can be ignored because they are typically an order of magnitude smaller than the reactance of the same element.

[16] Development is under way on devices that can be inserted in series with transmission lines to change their effective series impedance. Some of these devices are currently (2002) undergoing field testing. If successful, these devices will provide some level of control of the paths by which power will flow.

[17] "VARs don't travel well," G. Loehr.

$$P_{12} = V_1 * V_2 / X_{12} * \text{sine}\theta_{12}$$

STABILITY

Stability refers to the ability of the generators in a power system to operate in synchronism both under normal conditions and following disturbances. Three categories of instability are:

- Steady-state instability;
- Transient instability;
- Dynamic instability.

Steady-state instability refers to the condition where the equilibrium of the generators connected to the power system cannot accommodate increases in power requirements that occur relatively slowly or when a transmission line is removed from service for maintenance.

Since the power flow from one point to another is proportional to the sine of the angular difference between the voltages at the two points and inversely proportional to the total impedance of the circuits connecting the two points, there is a maximum level of power flow, that is, the delivery level at which the angular difference is 90 degrees, and the sine is equal to one.

The fact that the power flow is dependent on the sine of the angular difference between the voltages has an important significance in that it defines the maximum amount of power that can be moved across the facilities connected by the impedance X_{12}. If the power required by the customers at bus 2 is greater than the amount that can be delivered at a 90 degrees separation in the voltages, the system is unworkable. A technical term to describe a situation where the customer load at bus 2 slowly increases and the angular spread responds until it reaches the 90 degree point and then goes beyond 90 degrees is that the system becomes unstable and will collapse.

If the net impedance is increased by removing a line, less power can be transmitted. If there are a number of lines connecting bus 1 with bus 2, the loss or outage of any one of them will increase the impedance between the two buses and the system can again become unstable. Conversely, if the net impedance is reduced, more power can be transmitted. The value of the net impedance can be reduced by:

- Building an additional line(s) in parallel;
- Raising the design voltage of one or more of the existing lines;
- Decreasing the impedance of any of the existing lines by inserting a capacitor in series (remember X_C cancels out X_L).

Transient instability refers to the condition where there is a disturbance on the system that causes a disruption in the synchronism or balance of the system. The disturbance can be a number of types of varying degrees of severity:

- The opening of a transmission line increasing the X_L of the system.
- The occurrence of a fault decreasing voltage on the system. (The voltage at the fault goes to zero, decreasing all system voltages in the area.)
- The loss of a generator disturbing the energy balance and requiring an increase in the angular separation as other generators adjust to make up the lost energy.
- The loss of a large block of load in an exporting area.

When there is a disturbance on the system, the energy balance of generators is disturbed. Under normal conditions, the mechanical energy input to the generator equals the net electrical energy output plus losses in the conversion process within the turbine generator and the power consumed in the power plant.

If a generator sees the electric demand at its terminal in excess of its mechanical energy input it will tend to slow down as rotational energy is removed from its rotor to supply the new increased demand. If a generator sees an electric demand at its terminal less than its mechanical energy input, it will tend to speed up due to the sudden energy imbalance. This initial reaction is called the inertial response.

Disturbances may also change the voltage at the generator's terminals. In response, the generator's automatic voltage regulating system will sense the change and adjust the generator's field excitation, either up or down, to compensate.[18]

Transient stability or instability considers that period immediately after a disturbance, usually before the generator's governor and other control systems have a chance to operate. In all cases, the disturbance causes the generator angles to change automatically as they adjust to find a new stable operating point with respect to one another. In an unstable case, the angular separation between one generator or group of generators and another group keeps increasing. This type of instability happens so quickly, in a few seconds, that operator corrective action is impossible.

If stable conditions exist, the generator's speed governor system, sensing the beginning of change in speed, will then react to either admit more mechanical energy into the rotor to regain its speed or to reduce the energy input to reduce the speed. Directives may also be received by the generator from the company or area control center to adjust its scheduled output.

[18] See Chapter 8.

In addition to the measures noted to improve steady-state stability, other design measures available for selected disturbances to mitigate this type instability are:

- Improving the speed by which relays detect the fault and the speed by which circuit breakers operate to disconnect the faulted equipment sooner;
- The use of dynamic braking resistors which, in the event of a fault, are automatically connected to the system near generators to reduce export from the generators;
- The installation of fast-valving systems on turbines, allowing rapid reduction in the mechanical energy input to the turbine generator;
- Automatic generator tripping;
- Automatic load disconnection;
- Special transmission line tripping schemes.

Dynamic instability refers to a condition where the control systems of generators interreact in such a way as to produce oscillations between generators or groups of generators which increase in magnitude and result in instability, that is, there is insufficient damping of the oscillations. These conditions can occur either in normal operation or after a disturbance.

RESULTS OF INSTABILITY

In cases of instability, as the generator angles separate, the voltage and current angular relationships at points on the system change drastically. Some of the protective line relays will detect these changes and react as if they were due to fault conditions causing the opening of many transmission lines.

The resulting transmission system is usually segmented into two or more electrically isolated islands, some of which will have excess generation and some will be generation deficient. In excess generation pockets, the frequency will rise. In generation deficient pockets, the frequency will fall. If the frequency falls too far, generator auxiliary systems (motors, fans) will fail, causing generators to be automatically disconnected by their protective devices. Industry practice is to provide for situations where there is insufficient generation by installing under frequency-load-shedding relays. These relays, keyed to various levels of low frequency, will actuate the disconnection of blocks of customer load in an effort to restore the load-generation balance.

In situations where the frequency rises because of excess generation, generators will be automatically removed from service by protective devices detecting an overspeed condition. If studies indicate potential excess generation pockets, special, selective generation disconnection controls can be installed.

4

ELECTRIC ENERGY CONSUMPTION

This chapter discusses customer electric energy requirements. Industry practice is to refer to these requirements as "customer load". To supply customer load, the power system must also supply the losses in the system that delivers the energy. These losses are typically between 7% and 12% of the power delivered to consumers, larger than any single consumer. The power system planner must forecast and provide capacity to meet these needs.

END-USES FOR ELECTRICITY

Electricity serves a wide variety of end-uses. Each customer served by an electric utility has his own unique set of electric use requirements. Individual customer requirements for electric energy vary in magnitude and timing and reflect an almost endless array of devices employed individually or in combination with other devices. Among the principal categories of electric end-use equipment are:

- Electric motors;
- Electric furnaces;
- Lighting equipment;
- Space conditioning systems;

Understanding Electric Power Systems: An Overview of the Technology and the Marketplace, by Jack Casazza and Frank Delea
ISBN 0-471-44652-1

- Electric communications and computing devices;
- Machine tools; and
- Electrolytic devices.

Within each of these principal categories a wide variety of specific applications can be found. For example, electric motors include such sub-categories as compressors, pumps, fans, and conveyors. Furthermore, even within such sub-categories specific types of equipment may differ substantially in size, efficiency, and time pattern of electricity requirements.

CUSTOMER CLASSES

Customers of electric utilities also differ widely in their requirements for electricity. Such differences are reflected in both the timing and magnitude of requirements. Although each electric utility customer tends to have a somewhat unique pattern of electricity use requirements, customers are frequently segregated into classes of use for technical, administrative, and regulatory purposes. The primary classes of electric customers separate users into residential, commercial, industrial, and agricultural sectors. Some specific utilities, however, may identify other primary classes of customers. For example, generating utilities may have wholesale power customers. Customers in this class generally include other generating and non-generating utilities, which generally resell the purchased power to their own retail customers.

Other retail classes of customers may include street lighting systems, electric rail systems, churches, and governmental users of electricity. Still other customers are considered preference customers. These are typically found on federal- or state-owned utilities (Bonneville Power Administration, Southwest Power Administration, Tennessee Valley Authority, and Power Authority of the State of New York). These utility systems were developed specifically to provide electric service to these customers, who therefore must receive all of their electrical requirements at the lowest costs before power can be sold to other customers inside or outside the service areas of these systems.

Within these broad classes of service, customers may be further segregated on the basis of other technical and administrative parameters. Common distinctions of this type include:

- Phase and/or voltage level of service;
- Whether service is from underground or overhead facilities;
- The magnitude and/or timing of requirements;
- Type of specific end-use or major end-use requirement;
- Physical location of the end-use activity; and
- Quality or reliability of service provided.

Customer Class	Millions kWhrs in 2000	% of Total
Residential	1,192,446	34.9%
Commercial	1,055,232	30.8%
Industrial	1,064,239	31.1%
Other	109,496	3.2%
Total	3,421,414	100.0%

Figure 4.1. Customer electrical consumption 2000.‡‡‡
‡‡‡ Source—EIA Table 44, Electric Utility Sales by Sector—Includes energy service provider (power marketer) data.

In the United States, retail sales to the three major customer classifications are approximately equal in magnitude for the year 2000, as shown in Figure 4.1. While retail sales have grown 23.6% since 1991, the mix of its components has changed, with the commercial sector's share of the total growing from 27.4% in 1991 to 30.4% in 2000 and the industrial sector's share dropping form 34.3% to 31.4% over the same period.

RATE CLASSES

Such distinctions in the character of service provided are frequently the basis for the establishment of rate classes for regulatory and rate setting purposes. For example, within the residential sector separate rate classes are often established for customers with electric space heating, or electric water heat, requirements. Some electric utilities have special rate classes for all electric homes. Commercial and industrial customers may constitute separate rate classes or may be grouped together in general service rate classes. Also, within the broad category of general-service customers, separate rate schedules may be established on the basis of the magnitude of customer requirements, type of customer end-use, and the quality of service provided.

Definitions of rate classes may vary substantially among utilities and, between service jurisdictions within the same utility. A single utility, which provides service in more than one regulatory jurisdiction (e.g., more than one state), is likely to have a different set of rate class definitions in each jurisdiction.

Some states require that major electric systems provide daily kilowatt demand load curves for each electric consumer class for which there is a separate rate. This means that major electric consuming devices (such as air conditioners, water heaters, and space heating systems in residential dwellings) must be sub-metered for a sample of customers. For other load research activities, utilities will occasionally measure the consumption patterns of customers on a single transmission or distribution circuit from substation or transformer locations. These and other aggregate measures of consumption may also be used in the planning and operation of an individual utility's system, power pools, or reliability councils.

DEMAND AND ENERGY

The metering of electric utility service typically involves measurement of the amount of electricity used over a period of time and/or measurement of the rate (or rates) of use during the same period. Measures of the rate of use are measures of "demand", and they are usually stated in terms of watts or kilowatts (i.e., 1,000 watts). A customer who turns on a 100-watt light bulb creates an instantaneous demand of 100 watts on the electric system. If that light bulb were to operate continuously for one hour, 100 watt-hours of electricity would be consumed. The watt-hours of consumption represent the "energy" measure of the customer requirements. If ten 100-watt light bulbs were operated continuously for one hour, the customer would consume one thousand watt-hours, or one kilowatt-hour, of electric energy.

Consumer demands (i.e., the rates at which customers consume electricity) are not generally measured on an instantaneous basis. Instead, measures of customer's demands generally compute an average level of usage over a relatively short period of time. The most common periods for measuring customer demands are 15-minute, 30-minute, and one-hour intervals. However, in special circumstances, demand measures as short as one minute, or as long as three hours have been employed. A daily pattern of electricity use for a utility system is shown on Figure 4.2.

ENERGY

For billing purposes, electricity consumption by smaller customers (e.g., residential and small general-service customers) is typically measured only in terms of the kilowatt-hours consumed. Only rarely are the maximum demands of smaller users of electricity directly metered. As the magnitude of customer requirement increases, the importance of measuring the maximum rate of consumption by the customer also increases. Most utilities install meters which measure both kilowatt-hour consumption and maximum monthly demands for all customers that have a pre-established consumption level. For example, a utility may require demand metering for all customers whose monthly demands are expected to exceed 25 kilowatts and/or for all customers whose monthly consumption exceeds 6,000 kilowatt-hours. For certain very large cus-

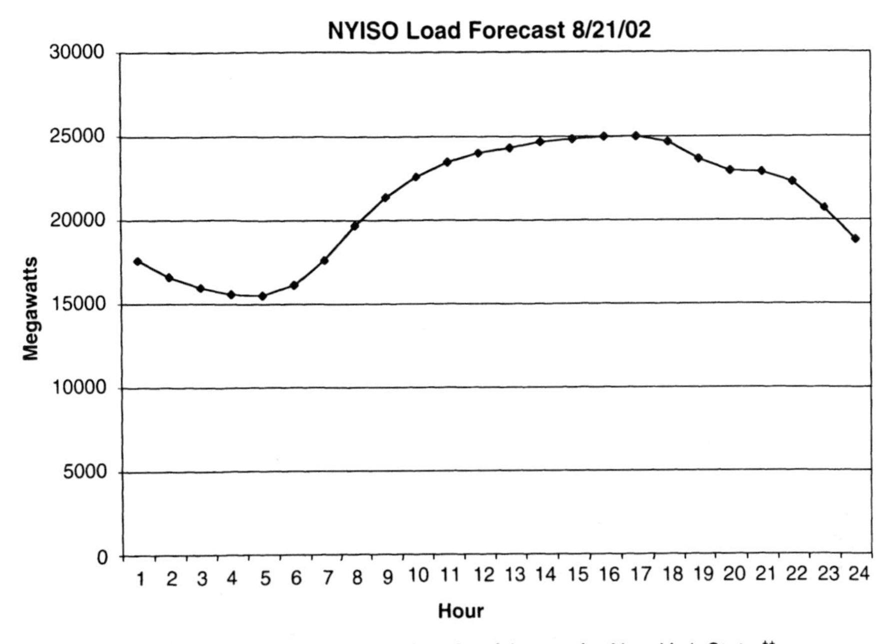

Figure 4.2. Daily pattern of summer weekday electricity use for New York State.[††]
[††] Source—NYISO Website.

tomers or classes of customers, demands may be measured on a continuous basis (as opposed to just monthly maximum) for every 15-minute, 30-minute, or hourly period throughout a month.

The recent trend toward time-of-use rates (i.e., the cost per kilowatt-hour varies at various times of the day) has also required the metering of energy consumption and/or maximum demand by time-of-use periods. These rates require metering of consumption for discontinuous periods within each billing month. For example, one utility defines its peak period for large general-service customers to include all weekday hours between 12 noon and 8:00 p.m.

EFFECTS OF LOAD DIVERSITY

There are two types of load diversity—that of different peak loads between customer classes, and that of different peak loads at different hours of the day and days of the year.

From continuously recorded demand data, a number of different demand measures may be derived. The most commonly metered measure of customer demand is individual customer maximum demand. This measure indicates the highest demand level incurred by the customer during any metering interval in a billing period. Due to varying types and uses of electrical equipment across customers, there are broad differences in the times that customers achieve their individual maximum demands. For example, one customer whose major use of electricity is outdoor light may regularly experience maximum demand in the late evening hours of the day; while a second customer, whose major requirement for electricity is for air conditioning, is more likely to experience a maximum demand during afternoon hours in the summer months. Furthermore, the electrical load requirements of an industrial process may be closely tied to work shift hours, or may be nearly flat throughout the 24-hour day if the process operates on a continuous basis.

These differences in the timing of individual maximum demands are referred to as diversity. Diversity in load requirements not only exists among individual customers, but also can be observed among rate classes, customer classes, jurisdictional divisions, utility systems, and power pools.

The inverse of diversity is coincidence. Coincident demand measures the maximum amount of load which occurs within a given measurement interval. If a customer has two or more electricity consuming devices at their facility, or residence, the customer's individual maximum demand will occur at the time at which the requirements of the individual devices are most highly coincident (i.e., demonstrate the least diversity). The sum of the maximum requirements of the individual devices will always be greater than, or equal to, the customer's individual maximum demand.

Other measures of individual customer demands tend to relate the individual customer's requirements to rate class, customer class, jurisdictional, or system requirements. Owing to diversity among customers, each individual customer's contribution to class, jurisdictional, or system maximum requirements, cannot be greater than, and tends to be less than; the customer's individual maximum demand.

System, jurisdictional, class, or customer demand are typically measured on an annual, monthly, or daily basis. However, billing data (both demand and energy measures) typically do not correspond directly to calendar month or calendar year measures. Usage reported on a bill in January, for example, may include substantial amounts of consumption which occurred in December of the previous year. This occurs because cost-effective meter reading and bill processing schedules generally require that not all meters be read on the same day but rather some are read on each working day of each month. Each customer is billed based on his metered consumption in the prior billing cycle. Measures of actual requirements on a calendar month basis are not available for individual customers, except where expensive continuously recording demand meters are installed.

At the system level, monthly aggregate requirements are usually determined by adding net electricity interchange into and out of an area to the net generation within the area from continuously recorded data typically maintained by utilities at the generation and transmission levels.

SYSTEM LOAD

The sum of each customer's coincident demand contributes to the "native" load of a utility system. This is the requirement for electric power placed on the utility's physical system by customers located within its franchised service area. It includes the actual customer requirements plus any electrical losses on the transmission and distribution system which are incurred when supplying these customer requirements. These losses are typically from 7% to 12% of customer energy consumption and as much as 50% of total system reactive requirements.

In addition to customer requirements and electrical losses, the electric utility's generating equipment must also provide electric power for the utility to operate its equipment. This includes the electricity needed to run various auxiliary devices within the generating stations (pumps, heaters, motors, etc.) and is known as "station service." This is typically about 5% of the energy produced in steam-generating plants, and less for gas turbines and hydro plants.

Net peak system demand is the load used for planning purposes. This includes customer loads and system losses, but not the "station service." Figure 4.3 is a typical annual load duration curve which shows that the peak load occurs for only a small percentage of hours each year. In many areas because of the weather sensitivity of customer demand, there are usually only a few percent of the hours in a year when the system load is at or greater than 95% of the peak load.

The net generation on a utility system includes net system load, plus or minus purchases or sales to other utilities. This load is sometimes called total load. When load data are obtained from utilities, it is essential that it be clearly delineated whether native load or total load is desired. It is also important that net loads be obtained since this is the industry norm.

The load experienced in any one area varies both daily, weekly and

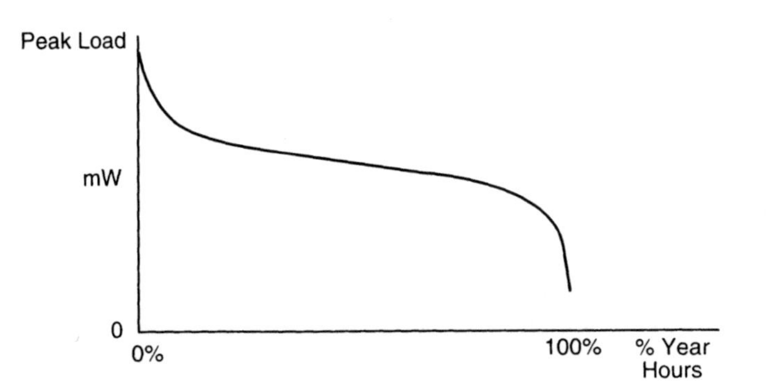

Figure 4.3. Annual load duration curve.

seasonally. This variation is measured by a quantity called a load factor. A load factor is defined as the ratio of average load to peak load during a specific period of time, expressed as a percent. Residential loads tend to have poorer load factors than industrial loads. The lower the load factor, the lower the full utilization of installed generating and distribution capacity.

The diversity in customer peak loads can result in situations where the time of the peak load in a specific geographic area may differ from the time of the system peak, that is, a residential area peaking in the evening while the system peaks during the late afternoon.

LOAD MANAGEMENT

Until the 1970s, emphasis was placed on ensuring that sufficient generating capacity was available to meet the peak load. That emphasis, however, was extremely expensive because of the rising cost of building generation plants, some of which only were needed for a few hours each year. Utilities instituted programs to manage demand so that the peak load could be reduced to obviate the need for the additional generating capacity.

Load can be reduced by the utility by charging more for power at certain times of the day or year, or it can be controlled directly by the electric utility (load management).[1] In load management, the electric company connects a control device to one or more appliances on the customer's premises. During periods of high demand, the electric company turns off the appliance for a period of time, thereby making it less necessary for the utility to keep generating reserves. In a sense, the utility has created additional interruptible customers. Tests indicate that, for many systems, the cost of load management equipment per KW of demand reduction is far less than that of capacity added to meet each KW of additional demand. However, the reduction in tariff required for customer acceptance may be quite large.

Although the exact definitions vary considerably, utility load management techniques are usually divided into two major classes: direct control, in which the utility controls end-use devices by means of supervisory control, and indirect control, for which load shifting is left to the discretion of the customer by provision of price incentives. It is not immediately clear how interruptible service contracts would be classified in such a partitioning. Some define two slightly different classes, namely: active load management, which permit the utility to control, interrupt, or displace a fairly precise quantity of load for a fairly precise time interval, to accommodate a fairly specific need and passive load management, which is subject to customer whim, and may or may not

[1] Over the longer range electric load can also be reduced by improvements in the design of more efficient lighting, motors and so on. Appliances such as refrigerators and air conditions have energy efficiency ratings.

result in an optimum benefit to the utility. In this case, the former class clearly includes interruptible service as an active load management technique.

For the purposes of this discussion, the two classes will be designated utility-controlled load management and customer-controlled load management. Utility-controlled load management shall include all load modification (or displacement) measures over which the utility retains complete control to implement or not, as the need arises. Customer-controlled measures, on the other hand, not only include measures over which the customer maintains complete control, but also fixed load control agreements (such as preset, timed switch on a water heater), which, from the moment to moment, are not under the direct control of either party.

A classification of various utility-controlled and customer-controlled load management practices is presented in Figure 4.4. As the figure shows, utility-controlled measures are divided into those such as storage and pooling, which tends to increase the amount of capacity during peak systems, and those such as interruptible service or remote load control, which attempt to reduce the load at these times. In either case, the net effect is to increase the difference between available generating capacity and system load during peak periods. All of the utility-controlled techniques listed may be used to reduce either daily or seasonal variations in system load.

Figure 4.4 also shows two classes of customer-controlled load management. The first of these includes all those measures taken by customers to meet their own end-use requirements for energy during utility peak periods. Such measures include self-generation and energy storage. The second class includes any customer-initiated actions to either deter or curtail specific end-use loads during utility-defined peak demand intervals. These actions may be based either on some negotiated contract or agreement between the customer and utility or on the customer's voluntary response to other incentives, such as time of day rates.

Utility Controlled		Customer Controlled	
Supply Side	**Demand Side**	**Backup or Storage**	**End-Use Modification**
Energy Storage	Interruptible Power	On-Peak Self Generation	Load Deferral
Power Pooling	Remote Control of Customer Load	Customer Energy Storage	Load Curtailment
			- Under contract
			- Voluntary response to incentives

Figure 4.4. Possible classification of utility load management techniques.

REACTIVE POWER

So far in this chapter we've talked about real power. As discussed in Chapter 3, there is another aspect of power, that is, reactive power, which is needed by equipment that involves the use of coils or motors, and so forth, requiring a magnetic field for proper operation. These devices look like an inductive load to the system and engineers refer to this effect as resulting in a lagging power factor on the system.

This reactive power can be provided from a number of sources, one of which is the generators on the electric power system. Other sources are capacitors connected between lines and the ground (called "shunt capacitors") and the capacitance effect of high-voltage cables and long-distance overhead transmission lines.

A power system will not function properly and will not remain in operation unless sufficient reactive power is available equal to the reactive loads plus the large reactive losses on the system.

FORECASTS

To a large extent, the business of the electric power industry is driven by forecasts. Forecasts of peak electric demand are used as the basis of long-range decisions for generation, transmission and distribution additions. While an aggregate forecast is usually suitable for generation planning purposes, the forecast has to be disaggregated for transmission and distribution planning purposes.

Forecasts of energy requirements are also used. These forecasts can be on a sendout basis or on a sales basis. The difference between sendout and sales is that sendout is the quantity measured at the output of the generators and is, in essence, "real time". Sales data, on the other hand, is that information obtained from the meters at customer's premises and is usually the total electricity used by the customer in the prior month (or "billing cycle—the period since the customer's meter was last read").

Sendout forecasts are used in the planning of generation resources since customer energy requirements determine the amount of fuel needed for generation. Sales forecasts are usually the basis of the revenue forecast needed for financial planning and rate making purposes.

Forecasts are usually prepared annually taking into consideration the most recent experience. Many participants base their forecasts on presumed weather conditions, that is, a temperature and humidity standard, based perhaps on the weather experience in a previous number of years. The experienced peak loads and energy sendout and sales are "weather normalized" and these values are used as the starting point for the forecasts.

The time horizon for peak load forecasts is driven by the planning cycle. There is no industry-wide magic horizon year for the forecast since the planning cycle differs form area to area, driven in large part by the time it takes to obtain state regulatory approval for new generation and transmission facilities. Generally speaking, in recent years, the lead time to construct new generation has decreased and the forecast horizon has also decreased. FERC, in its July 31, 2001, Notice of Proposed Rulemaking on Standard Market Design, indicates that Independent Transmission Service Providers will be required to forecast demand for a planning horizon of 3 to 5 years. Many consider these periods to be insufficient.

Energy and revenue forecasts tend to be driven by individual system and regulatory needs. In some states rate cases are based on forecast test years, requiring sales and revenue forecasts as inputs to a projected revenue requirement analysis.

Every forecast, since it depends on assumptions, has a degree of uncertainty. The longer the period the forecast covers, the greater the uncertainty. From a planning perspective, this creates a two edged sword—forecast too low and there may not be enough generation to supply customer load; forecast too high and revenue will be less than expected creating economic problems.

There are a number of approaches used to forecast peak demand. The most elementary being extrapolation of experienced loads. This technique assumes that the underlying customer usage pattern remains the same. This technique has applicability in short-term day-to-day and week-to-week forecasts for operating purposes.

Forecasts can be either "top down" or "bottom up". A top down forecast, called an econometric model, relates the future total system load to a set of macroeconomic assumptions. A variety of statistical modeling techniques are used to relate prior load to one or more explanatory variables. Various measures of employment, gross domestic product (GDP) values, and energy prices are among the more commonly used variables. Forecasts of these variables are made and used to drive the forecast. The forecasts can be based on a single model for an entire service area or there can be a number of models for different customer classifications, that is, residential, commercial, and so forth, each with its own explanatory variables.

A weakness of this approach is that, since it assumes that the underlying electrical uses remain the same as in the past; it does not capture the impact of new technologies or of significant changes in the underlying customer usage patterns. For example, a model based on 1970 and 1980 load data would not have captured the effect on aggregate load of the large influx of personal computers experienced in the 1990s. Models based on data from even the early 1990s wouldn't capture the trend for employees to work at home instead of in large office buildings.

A solution is to define the particular equipment of each customer class and its energy usage called an end-use approach. The forecast relates the future load to known or assumed changes in the stock of individual energy consuming devices. A residential customer might be identified by the number and type of its appliances, its cooling and heating systems, its lighting, and so on.

Using a residential end-use model the forecaster can explicitly address issues of turnover of appliances, appliance efficiencies, DSM impacts, new technology impacts and so on. This flexibility comes at a cost, that is, the large volume of information that is required. An aggregate consumption is developed for each customer and then for all the customers of the same type. This approach requires information on the types of appliances customers have, the electrical requirements for each including information about its age and efficiency (to factor into the forecast the effects of appliance turnover with newer, higher efficiency devices), the types of equipment they are buying, the average time period each is used and the coincidence factor of usage patterns between customers.

Commercial end-use modeling is more complicated than residential end-use modeling because of the complexity of the systems employed by these customers and the attendant problem of acquiring the necessary data about them. The Commend Program developed by EPRI defines twelve building types, ten end-uses and four fuel types (to allow for evaluation of the potential of substitution of alternate energy sources for electricity). The squared footage for each building type must be provided as well as an energy use per square foot. Changes in the customer base, for example, new office space coming into service, can be factored into the model.

Industrial customer end-use models are generally not used because of the variety of individual industries that comprise the sector, each one requiring its own model.

One approach, sometimes called the bottom-up approach, applies some of these techniques to smaller areas or regions, which are then combined to larger region totals. This bottom-up approach has the advantage of diversity of assumptions. Some forecasts may be high, some low with errors compensating. With the top-down forecasts, if a major assumption is in error the entire forecast is in error.

In some instances a combination of techniques is used, an end-use model for residential load, an econometric model for commercial load and an extrapolation of prior load adjusted for known changes (plant openings or closings) for industrial load. The modeling techniques selected are determined to a large extent by the size and types of the electric load, the availability of data and the resources available to the forecaster.

LOSSES AND UNACCOUNTED-FOR ENERGY IN THE DELIVERY SYSTEM

All the energy that leaves the generating stations is not reflected in the bills sent to customers. The difference is attributable to two issues. The first is called unaccounted-for energy. This energy is not metered by the local utility and is usually due to theft of service. In some underdeveloped countries this category can be as much as 50% of the energy generated. The second is the losses in the system directly related to the electric characteristics of the delivery system. They are an important consideration when selecting new electric power policies, when locating new generating plants, when deciding what generator to run to supply the next increment of load, when deciding on the voltage level and conductor sizes for new transmission and when deciding on the amount of voltage support to provide.

Losses occur in both lines and transformers. Line losses are directly related to the square of the value of the current (I^2R). The greater the amount of electricity the delivery system carries, and the greater the distance the greater the amount of energy lost as heat. Transformer losses are of two types: no-load loss and load loss. No-load losses are related to hysteresis and eddy-current loss in the transformer core and are independent of the current. Transformer load loss is related to I^2R. The no-load losses vary as the third to fifth power of the voltage and increase significantly when voltages are outside of design range.

Transformer manufacturers consider the amount of losses as one element of the requirements when designing a transformer. Typically, the cost of anticipated losses is a tradeoff with the capital cost to purchase the transformer. Losses can be reduced by increasing the size, and hence the cost, of a transformer. Rustebakke[2] reports that the total losses at rated transformer output amount to approximately 0.3–0.6% of the rated kiloVolt-Amperes of the unit. On most power systems load losses are 60% to 70% of total losses with transformer no load losses being from 30%–40%. In recent years, as the wholesale electric power market has been deregulated; new dispatch procedures and the increased flow of electricity on the bulk transmission system over longer distances without a commensurate increase in transmission capacity has caused an increase in transmission losses.

[2] Rustebakke, Homer M. 1983, *Electric Utility Systems and Practices, Fourth Edition*, Wiley, New York.

5

ELECTRIC POWER—GENERATION

Generation is the process of converting energy resources into electric power and energy, in order to be able to supply customer electricity requirements at all times. In describing the function of electric generating units, it is useful to view each unit as a complete system with input (fuel) and output (electrical energy). Between the input and output are various energy conversion devices—boilers, engines, turbines, electric generators—which convert energy contained in fuel to thermal energy (heat), thermal energy to mechanical energy and, finally, mechanical energy to electricity.

The generation of electric power is accomplished through the coordinated operation of several hundred physical components. The specific process depends on the types and sizes of plants. Some of the more general elements of electric power generation are best understood by reviewing the various types of plants and their characteristics.

In order to analyze, operate and plan generating systems, certain types of data are usually required, for example, efficiency data, such as heat rates, the amount of fuel required to produce a kilowatt-hour of electricity. This information, along with the fuel cost, is used in scheduling the division of load among the various generators on the system. For more than 80 years this was done to minimize total production costs. Since restructuring, this information is used by the plant owners to determine their prices and maximize their profits in the selling of bulk electric power.

The assurance of sufficient capacity is another requirement for a generating system. For this purpose, equipment breakdown data, outage data and

Understanding Electric Power Systems: An Overview of the Technology and the Marketplace, by Jack Casazza and Frank Delea
ISBN 0-471-44652-1

maintenance requirements are essential. Through the availability of such data, reliability analyses can be made to determine the adequacy of the total amount of generating capacity provided to meet the maximum demand that will be put on the system ay a given time.

It is also important that generating systems have sufficient fuel available or in storage for various types of future requirements if fuel supplies are disrupted for any reason. Data for this purpose, and for minimizing fuel costs, must be kept.

TYPES OF GENERATION

Many types of generating plants are in use and possible for the future, including steam plants fueled by coal, oil, or gas, nuclear plants, hydro plants and plants which use renewable energy. Most types of electric generating units can be grouped by prime mover—the type of device that drives the electric generator. The types of prime movers in use in the United States today are:

- Steam turbine;
- Combustion turbine; and
- Reciprocating engines.

Different fuels may be used for the various types of primer movers. The source of heat can be from the burning of coal, oil, gas or the heat given off in a nuclear reactor.

Steam Turbines

In a steam turbine generating plant fossil fuels (coal, oil, gas) are burned in a furnace. (In a nuclear plant heat is produced as a result of a nuclear chain reaction.) The heat given off by this combustion is used to heat water in a boiler to such a temperature that steam is produced. This steam (which may be as hot as 1,000 degrees F and at pressures as high as 3,600 psi) is then passed through one or more turbines. Energy contained in the steam is extracted by allowing the steam to expand and cool as it passes through the turbine(s). This energy turns the blades of the turbine, which are connected to a shaft. This shaft is connected to the electric generator and rotates the coils of the magnetic field of the generator, thus producing electricity. After passing through the turbine, the steam passes through a condenser, where it is cooled and becomes water for reintroduction to the boiler.

The functioning of a steam plant requires many pumps, fans, and auxiliary devices, particularly important are the feed water pumps, which force water through the boilers, the forced and induced draft fans, which provide sufficient air for the combustion in the boiler, and the system that injects the fuel into the boiler. After leaving the turbine additional heat is extracted from the steam in feed water heaters, heating water going to the boiler. The steam is then condensed and fed back into the boiler. Differences in the fuel used to produce heat will result in design and equipment differences for each generating station.

Combustion (Gas) Turbines

Combustion turbines are most often fueled by gas but can be fueled with some liquids. In a combustion turbine hot gasses (ignited fuel–air mixture) burn, are expanded through a turbine, driving a generator. An additional component of a combustion turbine is a compressor. This device increases the pressure of the air used in the combustion section by a factor of approximately 10. When the air is compressed in this manner, its temperature is increased. The resulting combustion of this heated air and fuel mixture raises the temperature of the gas to as much as 2,000°F. This gas is then passed through a turbine, where it is cooled and expanded. The dissipated energy turns the turbine, which, in turn, runs an electrical generator. Gas turbines are not as efficient as steam plants, but are considerably lower in capital costs. For this reason they are often used as "peaking" plants to supply peak electricity needs.

There are several variations on this basic design, each attempting to make maximum use of the energy input to the system. In some cases the turbine exhaust gases are used to preheat air prior to combustion. In other cases, the exhaust is used to heat steam in a boiler to operate in conjunction with a small steam turbine generator. This is known as a combined cycle plant. Combined cycle plants have excellent efficiency since a considerable amount of the energy in the gas turbine exhaust is recovered. Gas-fired combined cycle generation plants are currently the units of choice in the United States.

Hydro Turbines

Electric power is produced from water by directing a column of falling water past the "fins" of a hydraulic turbine. In a typical hydroelectric power plant, water is contained behind a dam. This dam causes the water level to rise. As a result, potential energy is stored in the water. To produce electricity, the water is made to flow through a turbine to a lower level. The difference in elevations between the two water levels is called the "head". In hydroelectric generation, the amount of electric energy that a given column of water is capable of producing varies directly with the head.

In the most common configuration, the hydraulic turbines and electrical generators are located at the site of the dam. However, it is not uncommon to have the turbines and generators located several miles downstream from the dam. Under this arrangement, water flows through a large pipe or "penstock" from the dam to the turbines.

There are several types of hydroelectric plants in use today. In the simplest form, a stream or river is diverted to pass through a hydraulic turbine. However, daily and seasonal variations in stream flow will cause the output of the hydro project to change. To deal with this situation, storage ponds are built. A portion of the stream flow is diverted into the storage facility during normal or high-flow conditions. Then, at low stream flow times, the water from the storage pond is released, thus maintaining the electric output of the project.

Pumped Storage

In still other configurations, water is stored in a lower reservoir and pumped to a higher elevation reservoir at night by using low-cost electrical energy produced by the utility's thermal plants at off peak times. During the peak load hours, the water is released from the upper reservoir and passed through a turbine or turbines, which drive electric generators. Often these turbine generators are reversible and used to pump the water in the lower reservoir back to the upper reservoir the next night. This type of hydroelectric station is known as pumped storage. On a net basis pumped storage plants do not produce electric energy but, in fact, consume energy, since about a third of the energy stored is lost in the process. They also perform a "relay" function. They receive low-cost energy produced at one time and place, hold it for a while, and then pass it on for use at another time in other places when alternate energy sources would be more expensive.

Nuclear Units

Nuclear units utilize a nuclear reaction as a source of heat for a conventional cycle. There are two primary designs used for nuclear reactors; boiling water reactors and pressurized water reactors. At the time of this writing, there are slightly more than 100 operating nuclear power units in the United States. No new nuclear units have been installed in the United States for many years because of concern over capital costs and safety. As nuclear units age, some owners are requesting extensions of their operation licenses. Other nuclear units are being retired.

Reciprocating Engines

This type of generation usually consists of a large diesel engine which uses #2 diesel fuel as a source of energy. Electricity is produced by connecting the output shaft of the engine to an electrical generator. Diesel engine improvements have resulted in considerable reductions in weight and improvement in efficiency.

Micro Turbines

These are small turbines that can be located at or near the customers. They can be installed on the customer's side of the meter or on the distribution or subtransmission system, depending on their size. They are usually fueled from the natural gas supply available.

Other Forms of Generation

There are other methods of producing electric power that currently contribute only small amounts to the total electric power production. There has been a renaissance, however, in the development of many new "small" technologies as fuel costs have increased. These technologies use sun, hydro, and other sources of energy to produce electricity. These methods include fuel cells, photo voltage cells, windmills, and so forth. A comparison of some of these small technologies is shown on the Figure 5.1.

Dispersed technology	Typical Size	Cost Compared to Conventional Plants			Use of Scarce Fuel Resources	Environmental Impact	Probable Date Of Introduction Into Power Systems	Special Considerations.
		Capital	Operating	Maintenance				
Small wind systems	100 W-750 kW	L	L	L	L	L	Now	An intermittent power source
Small scale hydro	1 kW – 15 mW-	L	L	L	L	L	Now	Output varies seasonally; seasonal pattern may vary from year to year
Solar								
Thermal	1-10 mW	G	L	G	L	L	5 – 10 yrs	Truly intermittent: operates only during daylight, output varies with weather
Photovoltaics	1–500 kW	G	L	L	L	L	5 – 10 yrs	Same as above, also requires very large surface area
Fuel cells	5kW-30 mW	E	L	G	Depends on fuel	L	5 – 15 yrs	Modular and expandable, no moving parts
Cogeneration	20-100 mW	G	L	L	Depends on fuel	Depends on fuel	now	Electrical output typically depends on demand for thermal output
Waste derived fuels	1–20 mW	G	L	G	Depends on emission controls	Depends on emission controls	now	Economics may require fuel cost to be "negative"
Storage								
Batteries	0–5 mW	E	L	L	L	L	Depends on technology	Battery lifea concern
Compressed Air	0–100 mW	G	E	E	L	E	now	Limited sites available
Microturbines	0-5 mW	G	E	G	G	L	now	Limited site availability
Diesels	0-10mW	E	E	E	G	G	now	Limited site availability

Figure 5.1. Comparison of small dispersed power production technologies.§§§
G = Greater than, L = less than, E = equal to
§§§ Some of the comparative data are derived from Cigre Paper #41-01

The interest in renewable energy sources has also spurred the development of improved methods for wind energy generating electricity. The wind is used to drive a turbine, which drives a generator. Due to the variable nature of the energy source, the speed of the wind turbine is not constant and machines other than synchronous generators have been applied. These include induction generators, dc generators, and variable reluctance generators. The further development of power electronic controls will doubtlessly result in new configurations for wind power generating systems. The effective generating capacity from wind power is considerably less than the capacity of the generators because there are periods when wind is not available.

Producing power directly from solar energy is not currently a significant factor in electricity generation. However, this technology has attracted the attention of electric utilities and industry as an alternative for future energy production. Photovoltaics are semiconductor devices that convert solar radiation (sunlight) directly into electricity. While the electricity is essentially free in that there are no fuel costs, photovoltaics have not been widely used, even in attractive climates due to the high initial investment required for the devices, their inability to produce power during periods of darkness, and the lack of knowledge about how long they will last. They are economic, however, for applications where expensive extensions of wires is needed to provide normal electric service, for example, railroad crossings.

Both wind power and solar power may need to be complimented with energy storage of some type to increase their effectiveness in supplying electricity requirements of the system.

CHARACTERISTICS OF GENERATING PLANTS

Some of the more common characteristics used to describe electric generating facilities are size, energy source, efficiency, type of use and availability. These elements are certainly not all inclusive of the information available to describe this equipment. In practice, hundreds, if not thousands, of measurements dealing with unit operation are made and/or recorded on a daily, hourly, or continuous basis. Many of these measurements describe the operation of the individual components of the generating facilities rather than the entire unit or plant.

Of overriding interest, however, are the characteristics of the overall generating plant consisting of a collection of fuel feeders, heat producers, energy converters, exciters, and electrical generators which must be operated in order for electricity to be produced. A plant may consist of several gas turbines, several boilers supplying steam to two or more steam turbines which drive two or more electric generators. A typical steam plant is shown in Figure 5.2, while a typical hydro plant is shown in Figure 5.3.

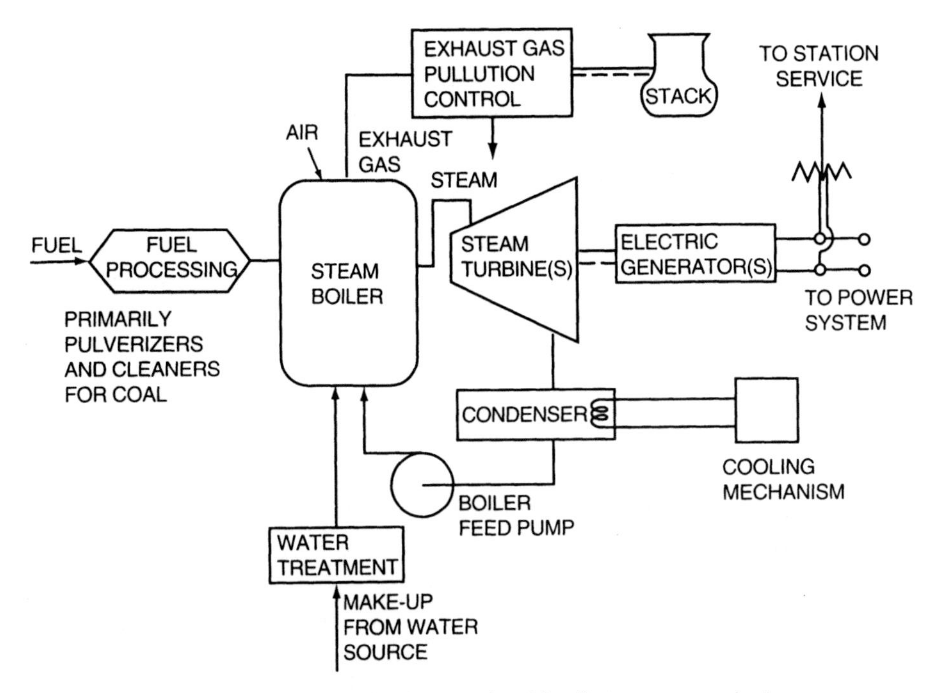

Figure 5.2. Schematic of conventional fossil steam power plant.

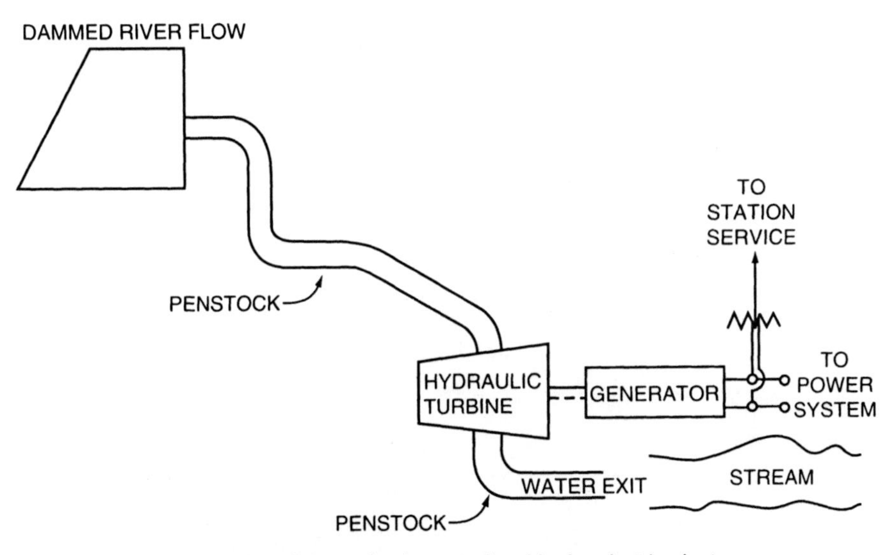

Figure 5.3. Schematic of conventional hydro electric plant.

Size

The size of a generating unit is measured by the number of megawatts it can produce for the electric power system. The capability of the unit may be limited by any of its components, for example, by the boiler, turbine, condenser, generator, or step-up transformer. Every generating unit comes from the manufacturer with a nameplate attached to it. This "plate" gives the designed capability of the electric generator and the steam turbine. This information is not usually used in determining a generating unit's actual capability. Both the mW and mVAr capabilities of generation are given in a capability curve, such as that shown in Figure 5.4.

Actual capability is determined by tests conducted on a regular basis where

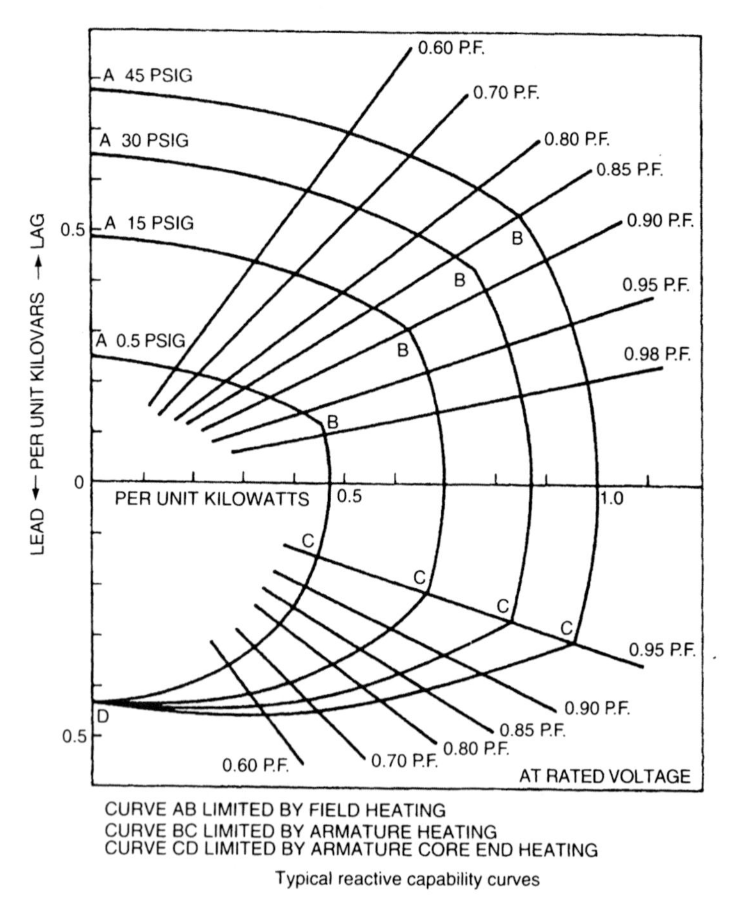

Figure 5.4. Generator reactive capability curve.****

**** From *Electric Utility Systems and Practices, 4th Edition*, Homer M. Rustebakke, Wiley, New York.

the output of a unit is demonstrated over a set time period. The results of these tests provide an indication of the maximum capability of the unit under normal and emergency conditions, that is, the ratings of the unit.

Depending on system conditions, such an ambient temperature, the duration of the production period, and the need for maximum capability, even at the expense of efficiency, generating units are given a number of different capability ratings which are usually different from the manufactures generator nameplate rating. Normal practice provides for the establishment of net dependable capabilities and emergency capabilities for generating units. Sometimes ratings are established for various seasons such as summer and winter, because of the effect of differences in ambient and cooling water temperatures. Net dependable capabilities are used for normal operation and are the basis for most production costing and fuel-use studies. Under emergency conditions, when extra capability is required to meet customer demand for electricity, various auxiliary devices such as feed water heaters may be bypassed, thereby decreasing unit efficiency but increasing its capability. Emergency capabilities are sometimes used for reliability calculations covering relatively infrequent occurrences of short duration.

Gross and net values may be referred to when discussing the size or capability of generating units. Gross capability is the output of a unit as measured at the terminals of the electric generator. However, many of the pieces of equipment used to operate the generating unit—crushers, pulverizers, pumps, building heaters, and so on—are themselves users of electric energy, typically about 5% of the unit's output in a steam plant. This electric energy is supplied by the generating unit through a special auxiliary transformer and reduces the amount of the unit's capability that can be delivered to the transmission system. This internal use of electric power is referred to as "station service" and the resulting capability that can be delivered by the plant is known as the net capability.

Generating units are built in sizes from a few kilowatts to over a thousand megawatts. The units driven by internal combustion engines and combustion turbines are generally at the lower end of the size scale. Gas driven microturbines are the smallest. Steam units have been built as small as several hundred kilowatts. Present day technologies and competitive economics are determining the best size of steam units. Figure 1.5 shows for the United States the number of units and the total capacity installed by fuel type.

The net dependable capability can change with changes in cooling water temperature, ambient air temperature, atmospheric pressure or the malfunction of a non-critical component. For example, a coal-fired generating unit has a number of pulverizers which grind the coal prior to combustion. The malfunction of one or more pulverizers could reduce the maximum fuel input with a resulting reduction in the capability of the unit. This is called a "partial outage."

The capability of a hydroelectric generating unit is determined by the size of the hydraulic turbine, the electric generator and the height of the water (head). The volume of water does not directly affect the maximum capability of the unit or plant, but rather affects the maximum amount of energy that may be generated in a given period of time. In practice, hydro units are generally scheduled to operate when they can replace the highest cost thermal generation. If the volume of available water is limited, the amount discharged per hour must be limited in order to permit the unit to operate for a desired period of time. Thus the capability of the unit can be effectively reduced for the period of low water conditions since the head will be reduced. When specifying the capability of hydro units, it is important to know the assumed water flow conditions on which the capability was based.

Efficiency

The thermal efficiency of a generating unit is a measure of the amount of electrical energy produced per unit of energy input. In the case of thermal units, the input is fuel and the means of measuring efficiency is called heat rate. Heat rate is defined as the ratio of the amount of fuel input (in BTUs) to a generating unit to the amount of electrical energy obtained (kilowatt-hours).[1] The resulting value is in units of BTU/kWh. For thermal units, this ratio will change as the output level of the unit is changed. In general, the efficiency of the unit will increase (i.e., the heat rate will decrease) as the output level of the unit is increased up to the normal rating of the unit.

A generating unit that is 100% efficient would use 3,412 BTUs of fuel to produce 1 kilowatt-hour of electric energy. Modern day steam units have heat rates in the 9,000–10,000 BTU/kWh range. Newer combined cycles have lower heat rates, some down to 7,000 BTU/kWh.

Dividing 3,412 by the unit's actual heat rate and then multiplying by 100 will give the overall percentage efficiency of a unit. When the heat rate is multiplied by the unit cost of BTUs in the fuel, the unit cost of the energy production cost ($/kilowatt hour) can be determined.

Two basic kinds of heat rates are used—average and incremental. The average heat rate simply defines the amount of fuel actually consumed for each hour that the generating unit is operating at a given level of electrical output. The incremental heat rate gives the increase in thermal input required to produce an additional increment of electrical energy, also in BTU/kWh. Changes as output decreases are sometimes called the decremental heat rates. Both the incremental heat rate and the average heat rate are given in units of BTU/kWh. Therefore, it is essential that the type of heat rate be defined when requesting or providing heat rate data.

[1] A BTU is the amount of heat required to raise the temperature of one pound of water one degree Fahrenheit.

Availability

The operation of a generating unit requires the coordinated operation of hundreds of individual components. Each component has a different level of importance to the overall operation of the generating unit. Failure of some pieces of equipment might cause little or no impairment in the operation of the unit. Still others might cause immediate and total shut-down of the unit if they fail. The failure rates of all the various components of a generating unit contribute to the overall unavailability of the unit. The unavailability of a generating unit due to component failure is known as its "forced outage rate". It is expressed as a percentage of time and is a measure of the amount of time the unit has been or might be unavailable to supply customer demand. The nature of forced outages is that they are random occurrences over which the plant owner has little or no control. The usual definition of forced outage rate is the number of hours that the unit is forced out of service divided by the sum of the number of hours the unit is connected to the power system plus the total number of hours that the unit is forced out of service.

There are two types of forced outages: partial and full. Partial forced outages are reductions in the capability of the unit due to failure of a component. Full forced outages occur when a critical component of the unit fails and the unit can no longer operate. This can happen in two ways: protective devices can "trip out" the unit, removing it from service; or, the plant operators can shut the unit down to protect equipment or personnel. Equivalent Forced Outage Rate (EFOR) is the term used to indicate the combination of full and prorated partial forced outage rates.

Another factor which contributes to unit availability is maintenance. Various components of generating units must be removed from service on a regular basis for preventive maintenance or to replace components before a forced outage results. Major maintenance would include turbine overhauls, generator rewinds and boiler retubing, for which complete shut-downs are required. In general, any condition requiring repair which can be postponed to a weekend is called a maintenance outage. If the unit must be removed from service during weekdays for a component problem, this is usually called a forced outage. Forced outages are events whose specific occurrence can not be predicted but which can be described by using probability measures. Maintenance outages are events which can be scheduled in advance. This difference is important in making analyses of total generator requirements for a system.

In the real world there are major areas of judgment and discretion involved in classifying availability data. They are often influenced by economic and reliability considerations. For this reason compilation and analyses of availability data requires extensive judgment and experience.

CAPITAL COST OF GENERATION

A key factor in selecting new generating facilities to be installed is their capital cost. The capital cost is the amount invested in the construction of the plant or any major improvements which might be made. During construction payments for equipment and contractors must be made. Those expenditures involve interest costs which must subsequently be recovered along with direct capital costs after the plant begins operating.

The combination of capital costs with those of fuel and operations and maintenance (O&M) determine the total cost of producing electricity at a plant. As previously discussed, this can be determined on an average basis or an incremental basis from both a short-term or long-term perspective. Short-term incremental costs would include fuel and labor. Long-term incremental costs would include fuel, labor, and maintenance. Both the return of capital and the return on capital must be considered in determining long-term costs.

TYPE OF USE

Generating units may be classified into three categories based on their mode of operation. These are:

1. Base Load;
2. Intermediate;
3. Peaking.

Base load units tend to be large units with low operating costs. They are generally operated at full capacity during most of the hours that they are available. They are designed to operate for long periods of time at or near their maximum dependable capability. Their low operating costs result from their use of low-cost nuclear and coal fuels and/or lower heat rates (higher efficiencies) than other units on the system. For a typical region, base load is on the order of 40% to 60% of the annual maximum hourly load and, since this represents the amount of load that will be supplied in the region at essentially all hours, it represents perhaps 60% to 70% of the annual energy requirements of the region.

Base load units are usually shut down for forced outages or maintenance only. Because of their size and complexity, these units may require from 24 hours to several days to be restarted from a "cold" condition. Once the decision has been made to shut down one of these units, periods of up to 24 hours may be required before another "start-up" may be attempted. When operating a power system decisions on the time of restarting units play an important role in hour-by-hour schedules for generation.

Intermediate units are those generating units which are used to respond to the variations in customer demand which occur during the day. They are designed to withstand repeated heating and cooling cycles caused by changes in output levels. Intermediate units usually have lower capital costs, and somewhat higher heat rates (lower efficiencies) than base load units. The intermediate load may be on the order of 30–50% of the maximum hourly load for a typical system and represents perhaps 20–30% of the annual energy requirements for the utility.

Peaking units are those generating units that are called upon to supply customer demand for electricity only during the peak load hours of a given period (day, month, year). Combustion turbines, reciprocating engines and small hydroelectric units comprise the majority of peaking units. These are ordinarily units with a low maximum capability (usually less than 150 Mw), which are capable of achieving full load operation from a cold condition within ten minutes. Peaking units usually have the highest heat rate sand lowest capital costs of the three categories of units. In addition to supplying system needs during peak load hours, they may be called upon to replace the capability of other base load or cycling units which have been suddenly removed from service due to forced outages. They generally supply about 5% of the total energy requirements of a system.

As generating units age, unit efficiency and performance generally decrease. In addition, newer, more efficient, lower operating cost units are continuously added to a power system. These two occurrences tend to cause most generating units to be operated fewer hours as they age.

LIFE EXTENSION

Most generating units were designed to have lives of 25–30 years. As units built in the past are retired, utilities must replace the capacity lost. In some cases the cost of performing major overhauls and modifications to those old units is more economic than building new capacity. The plant owners evaluate these options as part of their normal planning process.

The issue of life extension has become controversial because of the issue of enforcement of new source pollution standards. Under the Clean Air Act, existing power plants were exempt from the new standards. However, some older units have been upgraded so extensively in association with life extension programs that some maintain they are, essentially new units and, as such, should be subject to the standards.

SYNCHRONOUS GENERATORS

Essentially all of the world's electric power is generated by synchronous machines. The synchronous generator has proven to be a reliable and efficient device for converting mechanical power to electric power. Since the typical power system uses alternating current (60 Hz in the United States), the chief requirement of such a device is that it produces power at a controllable voltage at a constant frequency. A typical synchronous machine consists of a rotor with a field winding and a stator with a three-phase ac winding. The rotor has a dc power supply and the stator is connected to the power system through a generator step-up transformer. The turbine rotates the field at a constant speed, often as high as 3,600 rpm.

If the stator windings are connected to a load, current flows through the windings and the load. As the electrical load increases, the prime mover (turbine) must expend more mechanical energy to keep the rotor turning at a constant speed. Thus mechanical energy input by the turbine is being transformed into electrical energy. Generators in hydro plants are also synchronous machines, but rotate at lower speeds than steam unit or gas turbines.

The electrical power produced by a synchronous generator is almost equal to the mechanical power input, the efficiency being in the range of 98%. The division of electric load among a number of generators is determined by a number of factors, including economics. At a given operating point each turbine generator has an incremental cost, which is the cost per kWh to generate an additional small amount of power. Maximum system economy results when all generators are operating at the same incremental production cost.

The control of the real power and regulation of the speed (which must be held constant to provide a constant frequency) is done with the speed governor and automatic generator controls (AGC) and interaction with the system control center.

RESOURCE PROCUREMENT

Fuel provides the basic source of energy that is converted to electric power by electric generating units. The primary fuels used for the production of electric power are coal, oil, gas, nuclear fuel (uranium), and water. Solar energy and wind energy are presently contributing a negligible amount of the electric power produced in the United States today.

The physical units of measurement used to record a quantity of fuel consumed are tons, barrels and cubic feet for coal, oil, and natural gas, respectively. The given quantity of water used in hydroelectric power generation can produce varying amounts of electric power, depending on the distance through which it falls, or head, which will change as the water level behind the dam changes. Similarly, the quantity of nuclear fuel used is measured in terms of its total heat content or its ability to produce a certain amount of electric power—kilowatt hours.

Fuel arrives at generating stations from many different sources and by different modes of transportation. Even within a single utility system its generating stations are likely to receive fuel from different suppliers and regions of the country. These differences in source of supply and transportation cost contribute to large differences in the delivered price of fuels and their quality.

Fuel Measurements

One of the primary means of measuring the quality of fuel is to record the amount of heat contained in a given quantity. For fossil fuels (coil, oil, gas) this is usually given in units of BTU (British Thermal Units) per pound, gallon or cubic foot. The price paid by a utility for a quantity of fuel is usually keyed to the heat content of a particular shipment. For instance, coal arriving at two different generating stations from two different mines might have heat contents of 10,000 BTU per pound for one and 12,000 BTU per pound for the other. If both shipments were burned in identical generating units, one ton of the 12,000 BTU per pound coal would produce 20% more kilowatt hours of electricity than the 10,000 BTU per pound coal. It is for this reason that fuel prices are often quoted on a cost per BTU basis rather than a cost per ton, barrel, or cubit foot basis.

Other factors affecting the quality of fuels are the amount of non-combustible material contained in a given quantity. For coal and oil, most of these "impurities" are ash and sulfur. The amount of these elements contained in a shipment of fuel is determined by a chemical analysis conducted at the generating station and the results are given as percentage of the coal weight.

In quoting a price for a quantity of fuel, the supplier specifies the minimum heat content of the fuel as well as the maximum allowable quantity of other impurities. Any deviation from these specifications, as determined by the utility's chemical analyses, usually results in an adjustment of price paid for the shipment of fuel such that the quoted cost per BTU is maintained. Typical fuel prices are shown on Figure 5.5.

The delivery of natural gas to power plants is on a more continuous basis than either oil or coal deliveries. Consequently, measurement of fuel quality is less rigorous and on a more sporadic basis than for coal or oil.

Fuel	Sulfur Content	Price Range $/MMBTU
Oil:		
Heavy (No. 6)		1.50–3.50
Light (No.2)		
	Low Sulfur	1.35–1.50
	High Sulfur	3.00–4.50
Gas		3.00–6.00
Coal:		
	Low Sulfur	1.50–1.60
	High Sulfur	1.00–1.40

Figure 5.5. Typical fuel price ranges ($/MMBTU).

Fuel Transportation

The mode of transporting fuel from their source of supply to generating stations varies from station to station. Coal is generally transported from the mines to stations by rail. In some cases, due to the location of generating stations and regional geography, a portion of the coal transportation route is accomplished by barge over water. After delivery, the coal is deposited onto a coal storage pile. From there it is fed by conveyors into the hoppers of the generating stations. These hoppers are large bins designed to hold enough coal to run the generating unit for up to 24 hours between refills, depending on the output level of the unit.

Before being fed to the boiler, the coal passes through a series of grinders and pulverizers which reduce the coal to the consistency of talcum powder. These fine particles of coal are mixed with air and "sprayed" into the boiler and burned. Oil and natural gas generally arrive at the generating system by way of pipelines. Although some storage of oil is maintained, there is a closer correlation between delivery and use of these fuels than with coal. A large coal-fired power plant might receive one 80-car coal train per day and maintain as much as 90 days use in storage. Decisions have to be made between the amount of coal reserves maintained at the generating station to guard against interruptions in supply and the cost of maintaining this inventory. Oil tends to flow into a power plant on a intermittent basis with storage facilities of about 30 days. Gas is usually taken directly off of a pipe line with no storage at the power plant.

Fuel Used

Gas is the main source of fuel for new generation. It causes a low environmental impact and is currently readily available. A major concern, however, is the ability of gas pipelines to supply long-range future requirements and the ability of the gas industry to meet both electric system and local heating requirements.

Coal is the predominant fuel used for the generation of electric power in the United States. Use of coal requires compliance with environmental constraints given in the Clean Air Act and other legislation.[2] Coal is usually classified as high sulfur or low sulfur. The procedures to meet environmental requirements vary. Power plants have various devices such as precipitators and flue gas desulphurization systems to help. Another alternate that has been considered for meeting environmental requirements is the fluidized bed boiler.

The most common type of oil used in steam electric generating stations is residual or #6 oil. This grade of oil must be heated by burning higher (#2 oil) before it can be introduced into the boiler.

In addition to being used as a starting fuel and as a flame stabilizer (when

[2] See www.eia.doe.gov.

mixed with coal), #2 oil is also extensively used to run generators which are driven by diesel engines. The quantity of #2 oil used for these purposes is usually small enough to permit delivery by tank truck with a minimal amount of storage.

Some fossil-fueled power plants operating in the United States today have been designed with the capability of burning more than one type of fuel. This capability is usually designed into the combustion section or the furnace. Even if fuel-handling equipment or alternate fuel-storage capabilities do not currently exist at a power plant, this capability of burning an alternative fuel remains a valuable design feature. Over the past 15 years, the ability of burning alternate fuels in power plants has enabled utilities to mitigate the effects of long-term disruptions and response to significant shifts in the prices of fossil fuels.

Fuel Purchasing

Each generator may obtain fuel from several sources under a mix of arrangements. Some generators have long-term contracts for most of their fuel requirements with the price fluctuating over time but the quantity guaranteed. Other generators rely mainly on spot market purchase of fuels. Many generators use a combination of both procurement strategies to help ensure a continued supply of fuel at the lowest practicable price.

Emission Rights

Environmental legislation requires generators to have emission rights based on the type of fuel burned in the unit. Emission rights are traded in an open market, helping to lead to a minimum cost solution to meeting environmental requirements.

6

TECHNOLOGY OF THE ELECTRIC TRANSMISSION SYSTEM

This chapter discusses the elements of the transmission system. Included in the material are descriptions of overhead and underground transmission lines and substations. Transmission is the means by which large amounts of power are moved from generating stations, where this power is produced, to substations from which distribution facilities transport the power to customers. Transmission lines are also used to provide connections to neighboring systems.

COMPONENTS

The transmission system consists of three-phase transmission lines and their terminals—called substations or switching stations. Transmission lines can be either overhead, underground (cable) or submarine. There are high-voltage alternating current (HVAC) lines and high-voltage direct current lines (HVDC). Overhead transmission, subtransmission and primary distribution lines are strung between towers or poles. In urban settings underground cables are used primarily because of the impracticality of running overhead lines along city streets. While underground cables are more reliable than overhead lines (because they have less exposure to climatological conditions such as hurricanes, ice storms, tornadoes, etc.), they are also much more expensive than overhead lines to construct for unit of capacity and take much longer to repair because of the difficulty in finding the location of a cable failure and replacement.

Understanding Electric Power Systems: An Overview of the Technology and the Marketplace, by Jack Casazza and Frank Delea
ISBN 0-471-44652-1

HVAC

Overhead

The primary components of an overhead transmission line are:

- Conductors;
- Ground or shield wires;
- Insulators;
- Support Structures; and
- Land or right-of-way (R-O-W).

Conductors are the wires through which the electricity passes. Transmission wires are usually of the aluminum conductor steel reinforced (ACSR) type, made of stranded aluminum woven around a core of stranded steel which provides structural strength. When there are two or more of these wires per phase, they are called bundled conductors.

Ground or shield wires are wires strung from the top of one transmission tower to the next, over the transmission line. Their function is to shield the transmission line from lightning strokes. Insulators are made of materials which do not permit the flow of electricity. They are used to attach the energized conductors to the supporting structures which are grounded. The higher the voltage at which the line operates, the longer the insulator strings. In recent years, polymer insulators have become popular in place of the older, porcelain variety. They have the advantage of not shattering if struck by a projectile.

The most common form of support structure for transmission lines is a steel lattice tower, although wood H frames (so named because of their shape) are also used. In recent years, as concern about the visual impact of these structures has increased, tubular steel towers also have come into use. The primary purpose of the support structure is to maintain the electricity carrying conductors at a safe distance from ground and from each other. Higher-voltage transmission lines require greater distances between phases and from the conductors to ground than lower-voltage lines and therefore they require bigger towers. The clearance from ground of the transmission line is usually determined at the midpoint between two successive towers, at the low point of the catenary formed by the line.

RATINGS

The capability of an individual overhead transmission line, or its rating, is usually determined by the requirement that the line does not exceed code clearances with the ground. As power flows through the transmission line, heat is produced because of the I^2R effect. This heat will cause an expansion of the metal in the conductor and a resultant increase in the amount of its sag. The amount of sag will also be impacted by the ambient temperature, wind-speed and sunlight conditions. Ratings are usually of two types—normal and emergency—and are determined for both summer and winter conditions. Some companies average the summer and winter ratings for the fall and spring. In recent years there has been a trend to calculating ratings for critical transmission lines on a real-time basis reflecting actual ambient temperatures as well as the recent loading (and therefore heating) pattern.

Ratings are also specified for various time periods. A normal rating is that level of power flow that the line can carry continuously. An emergency rating is that level of power flow the line can carry for a various periods of time, for example, 15 minutes, 30 minutes, 2 hours, 4 hours, 24 hours, and so forth.

The land that the tower line transverses is called the right-of-way (ROW). To maintain adequate clearances, as the transmission voltage increases, R-O-W widths also increase. In areas where it is difficult to obtain R-O-Ws, utilities design their towers to carry multiple circuits. In many areas of the country it is not uncommon to see a structure supporting two transmission lines and one or more subtransmission or distribution lines.

There are different philosophies on the selection of R-O-Ws. One philosophy is to try to site the corridor where there is little if any visual impact to most people. The other is that the R-O-W should be adjacent to existing infrastructure, that is, a railroad, highways, natural gas pipelines, so as to minimize the overall number of corridors dedicated to infrastructure needs. Reliability concerns argue for as much separation as possible between transmission R-O-Ws, to minimize exposure to incidents which might damage all lines on a R-O-W, so called common mode failures, such as ice storms, hurricanes, tornadoes, forest fires, airplane crashes, and the like. An ongoing issue with R-O-Ws is that they must be maintained to avoid excessive vegetation growth, which reduces the clearances between the line and ground.

CABLE

The majority of the transmission cable systems in the United States are high-pressure fluid filled (HPFF) or high-pressure liquid filled (HPLF) pipe type cable systems. Each phase of a high-voltage power cable usually consists of stranded copper wire with oil-impregnated paper insulation. All three phases are enclosed in a steel pipe. The insulation is maintained by constantly applying a hydraulic pressure through an external oil adjustment tank to compensate for any expansion or shrinkage of the cable caused by temperature variations.

In recent years laminated paper-polypropylene insulation has been introduced. Also, cross-linked polyethylene insulated cable (XLPE) has come into use at lower transmission voltages. Since it has no metallic sheath, it has greater flexibility and lower weight, as compared with conventional paper insulated cable. Both oil-filled and XLPE cables are available today for operation at voltages up to 500 kV ac, although there is limited experience with XLPE cables operating above 220 kV.

Cable capacity is determined by concerns over the effect of heat on the cable insulation. Since the cable is in a pipe which is buried in a trench, dissipation of the I^2R heat is a major issue in cable design and operation. Cable capacity can be increased by surrounding the pipe with a thermal sand, which helps dissipate heat. To maintain integrity of the insulation, the splicing of cable sections is done under controlled environmental conditions. To increase the HPLF cable's capacity, the dielectric fluid can be circulated to mitigate local hot spots and to remove generated heat; air-cooled heat exchangers can be added to improve the dissipation of the generated heat from the circulated oil. At least one utility has employed a refrigeration cooled heat exchanger.

A limitation on the application of HPLF ac cables is their high level of capacitance, causing high charging currents, which limits the length of cable that can be used without some intermediate location where shunt reactor compensation can be installed. XLPE cables, since they do not have a metal sheath, have lower levels of capacitance.

Submarine Cables

Submarine cables are usually laid underwater in trenches, with the distance between each phase measured in feet. A major consideration is to have the trench deep enough and wide enough so that the cables are not damaged by anchors or fishing trawlers. The environmental impacts of dielectric fluid leaks from damaged cables are a concern. Also of concern is the need for long lengths of spare cable to facilitate repairs in the event of damage or failure. In any case, repair times will be long, possibly a month or months.

SUBSTATIONS

Substations are locations where transmission lines are tied together. They fulfill a number of functions.

- They allow power from different generating stations to be fed into the main transmission corridors.
- They provide a terminus for interconnections with other systems.
- They provide a location where transformers can be connected to feed power into the subtransmission or distribution systems.
- They allow transmission lines to be segmented to provide a degree of redundancy in the transmission paths.
- They provide a location where compensation devices such as shunt or series reactors or capacitors can be connected to the transmission system.
- They provide a location where transmission lines can be de-energized, either for maintenance or because of an electrical malfunction involving the line.
- They provide a location for protection, control, and metering equipment.

Substation Equipment

There are a number of designs used for substations. However, there are elements common to all:

- *Bus* is the given name given to the electrical structure to which all lines and transformers are connected. Buses are of two generic types: open air and enclosed. Enclosed buses are used when substations are located in buildings or outdoors where space is at a premium. They involve the use of an insulating gas such as sulfur hexafluoride (SF_6) to allow reduced spacing between energized phases. Bus structures are designed to withstand the large mechanical forces that can result from fields produced by high short-circuit currents. These forces vary with the third power of the current. A bus section is the part of a bus to which a single line or transformer is connected.
- *Protective relays* are devices that continuously monitor the voltages and currents associated with the line and its terminals to detect failures or malfunctions in the line/equipment. Such failures are called faults and involve contact between phases or between one or more phases and ground.[1] The relays actuate circuit breakers.
- *Circuit breakers* are devices that are capable of interrupting the flow of electricity to isolate either a line or a transformer. They do so by opening the circuit and extinguishing the arc that forms using a variety of technologies such as oil, vacuum, air blast or sulfur hexafluoride (SF_6). Breakers may be in series with the line or transformer or may be installed on both sides of the bus section where the line connects. They allow individual lines or transformers to be removed from service (de-energized) automatically when equipment (protective relays) detects operating conditions outside a safe range. They must be capable of interrupting the very high currents that occur during fault conditions and are rated by the amount of current they can interrupt. These fault current levels can be 20 or 30 times larger than the current flow under normal operating conditions, that is, thousands of amperes. To minimize the impact of electrical "shocks" to the transmission system, minimizing the total time for the relay to detect the condition and the circuit breaker to open the circuit is a critical design issue. Circuit breakers also allow lines or transformers to be removed from service for maintenance. Circuit breakers normally interrupt all three phases simultaneously, although in certain special applications, single-phase circuit breakers can be employed, which will open only the phase with a problem.
- *Transformers* are devices that are used to connect facilities operating at two different voltage levels. For example a transformer would be used to connect a 138 kV bus to a 13 kV bus. The transformer connects to all three

[1] A malfunction can also be a situation where one phase is open without contacting ground.

phases of the bus. Physically the transformers can include all three phases within one tank or there can be three separate tanks, one per phase. Larger capacity units may have three separate tanks because their size and weight may be a limiting factor because of transportation issues.

Transformers can be designed with two mechanisms to adjust the voltage ratio. One mechanism is the provision of more than one fixed tap position on one side of the transformer. For example, a transformer might have a nominal turns ratio of 345/138, with fixed taps on the 345 kV winding of 327.8, 336.7, 345, 353.6 and 362.3. The transformer must be deenergized to adjust the fixed tap ratio. Another mechanism is called tap changing under load (TCUL). In this mechanism the ratio can be adjusted while the transformer is energized, providing greater operating flexibility. Some transformers have both types of mechanisms; with a fixed tap adjustment in the high voltage winding and the TCUL adjustment in the low voltage winding.

Another type transformer is an *autotransformer*, which is used when facilities at nearly the same voltage are to be connected, for example, 138 kV to 115 kV. Rather than having two separate paths for the electricity, connected only by the magnetic flux through the transformer as in a conventional unit, the winding of autotransformer involves a tap on the higher voltage winding which supplies the lower voltage.

All *larger transformers* have mechanisms to remove the heat generated within the tank involving some manner of circulating the transformer insulating/cooling oil through an external heat exchanger involving fins mounted on the side of the transformer and fans to circulate air across the fins to maximize heat dissipation.

- *Disconnect switches* are used to open a circuit when only "charging" current present is due. These would be used primarily to connect or disconnect circuit breakers or transformers which are not carrying load current. They are also used in conjunction with circuit breakers to provide another level of safety for workers by inserting a second opening between station equipment out of service for work and the still energized section of line or bus.
- *Lightning arrestors* are used to protect transformers and switchgear from the effects of high voltage due to lightning stroke or a switching operation. They are designed to flashover when the voltage at the transformer exceeds a pre-selected level which is chosen by the station design engineers to coordinate with the basic insulation level of the transformer (BIL).
- *Metering equipment* is provided to measure line and transformer loadings and bus voltages so operating personnel can ensure that these facilities are within acceptable limits. Metering equipment also is provided at some locations to measure the flow of energy for the billing that is required for sales and purchases of energy between various participants in the electric energy market.

- *SCADA* is an acronym for system control and data acquisition. It reflects the improvements in measurement, telecommunications and computing technologies that allow more and more automation of substation operation.

Depending on the electrical characteristics of a particular part of the transmission system, other equipment that may be located at a substation are:

- *Shunt reactors* (reactors connected from the energized bus to ground) are installed to control high voltages that occur especially at night due to the capacitive effect of lightly loaded transmission lines. These reactors can be energized always or they can be energized only at specific times. Shunt reactors are also used to reduce or control the high voltages that can occur when a sudden loss of a block of customer load occurs. The windings, insulation and the external tank are similar to those used for transformers.
- *Series reactors* are installed in a transmission line to increase the impedance of the line, to decrease current levels in the event of short circuits, or to reduce its loading under various operating conditions.
- *Shunt capacitors* are installed to provide mVArs to the system to help support voltage levels.
- *Series capacitors* are installed to reduce the effective impedance of a transmission line. These would be installed in very long transmission lines to effectively reduce the electrical angle between the sending and the receiving parts of the system, enabling more power to flow over the line and increasing stability limits.
- *Phase angle regulating transformers* are installed to control power flow through a transmission line, causing more or less power to flow over desired lines. They use a variant on the design of a normal transformer, in which, due to the specialized way they are wound, they electrically inject an angular phase shift into the line. The angle can be made to either increase or decrease power flow on the line. Since they are expensive, they are often used on cable systems where, because of cost and limited capacity of cables, maximum utilization of all parallel cable capacity was essential. In recent years, some of them are being installed in overhead transmission lines to control parallel path flow, when power flows over paths in other systems not involved in transactions, or do not have adequate capacity.
- *FACTS* (Flexible ac Transmission Systems) is a generic name used for a variety of devices intended to dynamically control voltage, impedance or phase angle of HVAC lines. The development of such devices was first patented in 1975 by J.A. Casazza.[2] The development of such devices was

[2] Casazza, J.A., March 2, 1976, Power Injector Control Means for Transmission Service, U.S. Patent No., 3,942,032.

encouraged in the 1980s by a program of the Electric Power Research Institute (EPRI).[3]

These devices mirror and extend the benefits of the fixed series and shunt inductors and capacitors previously discussed in that the FACTS devices allow rapid and precise adjustments. Depending on the device, these FACTS devices provide a number of benefits: increased power transfer capability, rapid voltage control, improved system stability, and mitigation of sub-synchronous resonance (a condition experienced in a number of regions in the United States, where oscillations occur caused by interaction of generator control systems and the capacitance of long transmission distances). There are many devices by many manufacturers, some of which are in the development stage and a few of which are in service. The names of the devices vary somewhat, depending on the manufacturer. The following lists some of the devices:

- Static VAr Compensators (SVCs)—These devices employ fixed banks of capacitors, controlled with thyristors, which can switch them on and off rapidly. In many instances, there are also thyristor-switched inductors to prevent system resonance.
- Thyristor Controlled Series Compensators (or Series Capacitors) (TCSCs)—A thyristor controlled reactor is placed in parallel with a series capacitor, allowing a continuous and rapidly variable series compensation system.
- *Static Compensators* (STATCOMs) are gate turn-off type thyristors (GTO) based SVCs. They are solid-state synchronous voltage generators that consist of multi-pulsed, voltage sourced inverters connected in shunt with transmission lines. They do not require capacitor banks and shunt reactors but rely on electronic processing of voltage and current waveforms to provide inductive or capacitive reactive power. They have the added advantage that their output is not seriously impacted by low system voltage.
- *Unified Power Flow Controllers* (UPFC)—These devices have shunt connected STATCOM with an additional series branch in the transmission line supplied by the STATCOM's dc circuit. These devices are comparable to phase shifting transformers. They can control all three basic power transfer parameters: voltage, impedance and phase angle.
- *SVC Light*[4] (STATCOM)—Are based on voltage source converter technology equipped with Insulated Gate Bipolar Transistors (IGBT) a power

[3] FACTS—Flexible Alternating Transmission Systems for Cost Effective and Reliable Transmission of Electric Energy, K. Habur, Siemens AG, and D. O'Leary, World Bank; also derived from Westinghouse Flexible AC Transmission Systems (FACTS)—www.ece.umr.edu, and ABB FACTS—Flexible AC Transmission Systems—www.abb.com

[4] The ABB brand name.

switching component. They provide reactive power as well as absorption purely by means of electronic processing of voltage and current waveforms.

Substation Breaker Arrangements

Figure 6.1 shows the most commonly used bus/circuit breaker arrangements. The breaker and half design is the one most usually used in newer transmission substations since it provides excellent reliability and operating flexibility.

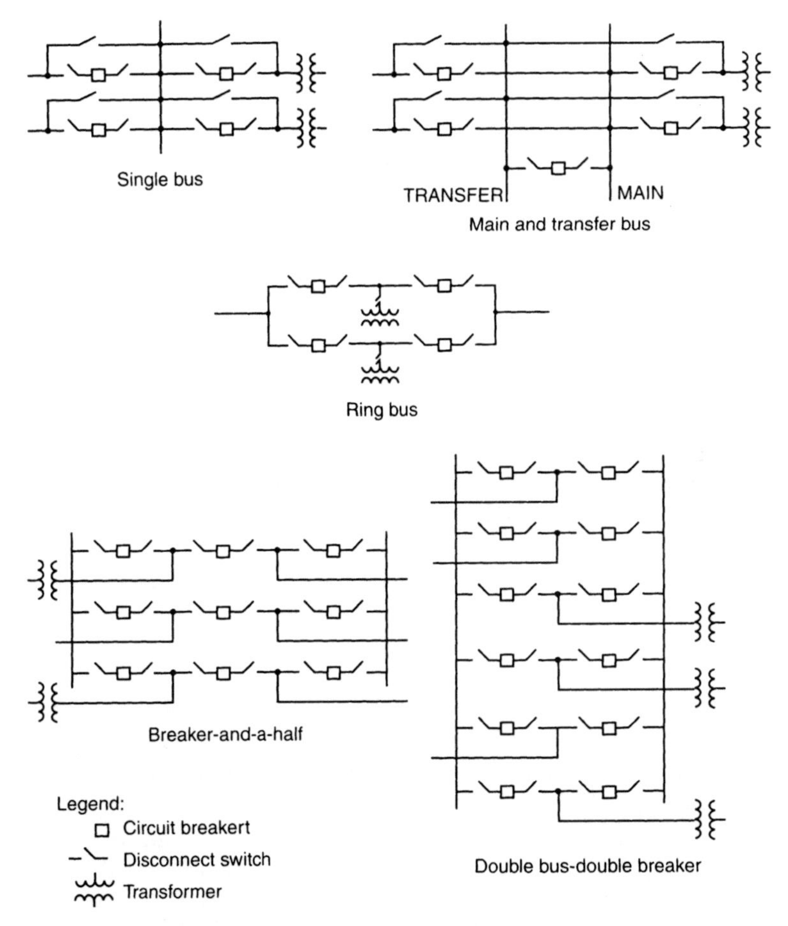

Figure 6.1. Typical substation circuit breaker arrangements.[††††]

[††††] From *Electric Utility Systems and Practices, 4th Edition*, Homer M. Rustebakke, Wiley, New York.

TRANSMISSION SYSTEM AGING

Transmission systems are aging with a large portion of the lines, cables, and substation equipment reaching average ages in excess of 30 years, and some in excess of 75 years. This has increased future failure rates and maintenance requirements[5] causing reliability problems.

HVDC

An alternate means of transmitting electricity is to use high-voltage direct current (HVDC) technology. As the name implies, HVDC uses direct current to transmit power. Direct current facilities are connected to HVAC systems by means of rectifiers, which convert alternating current to direct current, and inverters, which convert direct current to alternating current. Early applications used mercury arc valves for the rectifiers and inverters but, starting in the 1970s, thyristors became the valve type of choice.

Thyristors are controllable semiconductors that can carry very high currents and can block very high voltages. They are connected is series to form a thyristor valve, which allows electricity to flow during the positive half of the alternating current voltage cycle but not during the negative half. Since all three phases of the HVAC system are connected to the valves, the resultant voltage is unidirectional but with some residual oscillation. Smoothing reactors are provided to dampen this oscillation.

HVDC transmission lines can either be single pole or bipolar, although most are bipolar, that is, they use two conductors operating at different polarities such as +/–500kV. HVDC submarine cables are either of the solid type with oil-impregnated paper insulation or of the self-contained oil-filled type. New applications also use cables with extruded insulation, cross-linked polyethylene.

Although synchronous HVAC transmission is normally preferred because of its flexibility, historically there have been a number of applications where HVDC technology has advantages:

- The need to transmit large amounts of power (>500mW) over very long distances (>500km), where the large electrical angle across long HVAC transmission lines (due to their impedances) would result in an unstable system. Examples of this application are the 1,800mW Nelson River Project, where the transmission delivers the power to Winnipeg, Canada, approximately 930km away; the 3,000mW system from the Three Gorges project to Shanghai in China, approximately 1,000km distant; and the 1,456km long, 1,920mW line from the Cabora Bassa project in Mozambique to Apollo, in South Africa. In the United States the 3,100

[5] See Chapter 9 on Reliability.

mW Pacific HVDC Intertie (PDCI) connects the Pacific Northwest (Celilo Converter Station) with the Los Angeles area (Sylmar Converter Station) by a 1,361 km line.

- The need to transmit power across long distances of water, where there is no method of providing the intermediate voltage compensation that HVAC requires. An example is the 64 km Moyle interconnector, from Northern Ireland to Scotland.
- When HVAC interties would not have enough capacity to withstand the electrical swings that would occur between two systems. An example is the ties from Hydro Quebec to the United States.
- The need to connect two existing systems in an asynchronous manner to prevent losses of a block of generation in one system from causing transmission overloads in the other system if connected with HVAC. An example is the HVDC ties between Texas and the other regional systems.
- Connection of electrical systems that operate at different frequencies. These applications are referred to as back-to-back ties. An example is HVDC ties between England and France.
- Provision of isolation from short-circuit contributors from adjacent systems since dc does not transmit short-circuit currents from one system to another.

With the deregulation of the wholesale power market in the United States, there is increasing interest in the use of HVDC technology to facilitate the new markets. HVDC provides direct control of the power flow and is therefore a better way for providing contractual transmission services. Some have suggested that dividing the large synchronous areas in the United States into smaller areas interconnected by HVDC will eliminate coordination problems between regions, will provide better local control, and will reduce short-circuit duties, significantly reducing costs.

Advantages of HVDC

As the technology has developed, the breakeven distance for HVDC versus HVAC transmission lines has decreased. Some studies indicate a breakeven distance of 60 km using modern HVDC technology. Some of the advantages identified are:

- No technical limits in transmitted distance; increasing losses provide an economic limit;
- Very fast control of power flow, which allows improvements in system stability;
- The direction of power flow can be changed very quickly (bi-directionality);

- An HVDC link does not increase the short-circuit currents at the connecting points. This means that it will not be necessary to change the circuit breakers in the existing network;
- HVDC can carry more power than HVAC for a given size of conductor;
- The need for ROW is much smaller for HVDC than for HVAC, for the same transmitted power.

Disadvantages of HVDC

The primary disadvantages of HVDC are its higher costs and that it remains a technology that can only be applied in point-to-point applications because of the lack of an economic and reliable HVDC circuit breaker. The lack of an HVDC circuit breaker reflects the technological problem that a direct current system does not have a point where its voltage is zero as in an alternating current system. An HVAC circuit breaker utilizes this characteristic when it opens an HVAC circuit.

KNOWLEDGE REQUIRED OF TRANSMISSION SYSTEM

Those familiar with transmission system problems and policies have developed the following list, sometimes called the "ten commandments" of transmission knowledge:

Thou shall understand and consider:

1. How systems are planned and operated;
2. Effect of generation on transmission and vise versa;
3. Causes of circulating power, parallel path flow, and loop flow;
4. Differences between individual circuit capacities and transmission capacities;
5. Synchronous ac connection advantages and disadvantages;
6. Reactive power and its role;
7. Causes and consequences of blackouts;
8. Need for new technology;
9. Disincentives to building new transmission;
10. Need for special training and education.

Some of these have been discussed in this and previous chapters. The rest will be covered in the following chapters.

7

DISTRIBUTION

The primary function of the distribution system is to connect the electric bulk power system to customers requiring service at voltages below that of the transmission and subtransmission systems. The distribution system is the portion of the electric power system most readily seen by the customer and which contributes most directly to providing electric service.

Of the three primary functions of the electric utility, generation, transmission, and distribution, the distribution system plays the largest role in the quality of service received by the consumers. Figure 2.4 shows the relationship of the distribution system to the total system.[1] The primary components of a distribution system are:

- Distribution Substation;
- Primary Feeder;
- Distribution Transformer;
- Secondaries and Services.

The distribution substation receives electric power directly from the transmission or sub transmission system and converts it to a lower voltage for use on a primary distribution feeder. In a common configuration a distribution

[1] *Wiley Encyclopedia of Energy Technology and the Environment*, New York, February 1995, p. 459.

Understanding Electric Power Systems: An Overview of the Technology and the Marketplace, by Jack Casazza and Frank Delea
ISBN 0-471-44652-1

substation may have several transformers and a number of primary distribution feeders emanating from it. These feeders are most commonly seen being supported by wooden utility poles on residential streets.

The distribution transformer, usually on a pole, is supplied by the primary distribution feeder (primaries) and transforms the voltage of the primary feeder (2400 volts through 34,500 volts) to a lower voltage most commonly used by consumers. The secondary lines (secondaries) and service connections provide electric service directly to the ultimate consumer at the lower voltages produced at the output terminals of the distribution transformers.

PRIMARY FEEDERS

Primary voltage in the "13 kV class" is predominate among United States utilities. The 4-kV class primary systems are older and are gradually being replaced. In some cases 34 kV is used in new, high-density-load areas.

The three-phase, four-wire primary system is the most widely used. Under balanced operating conditions, the voltages of each phase are equal in magnitude and 120° out of phase with each of the other two phases. The fourth wire in these Y-connected systems is used as a neutral for the primaries, or as a common neutral when both primaries and secondaries are present. The common neutral is also grounded at frequent intervals along the primary feeder, at distribution transformers, and at customers' service entrances.

Rural and suburban areas are usually served by overhead primary lines, with distribution transformers, fuses, switches, and other equipment mounted on poles. Urban areas with high-density loads are served by underground cable systems, with distribution transformers and switchgear installed in underground vaults or in ground-level cabinets. There is also an increasing trend toward underground single-phase primaries serving residential areas. Underground cable systems are highly reliable and unaffected by weather, but can have longer repair times. The costs of underground distribution are significantly higher than overhead. Primary distribution includes three basic types: (1) radial, (2) loop, (3) and primary network systems.

Radial Systems

The radial system is a widely used, economical system often found in low-load density areas. To reduce the duration of interruptions, overhead feeders can be protected by automatic reclosing devices located at the substation or at various locations on the feeder. These devices reenergize the feeder if the fault is temporary. To further reduce the duration and extent of customer interruptions, sectionalizing fuses are installed on branches of radial feeders allowing unaffected portions of a feeder to remain in service.

Loop Systems

The loop system is used where a higher level of service reliability is desired. Two feeders form a closed loop, open at one point, so that load can be transferred from one feeder to another in the event of an outage of one circuit by closing the open point and opening at another location. One or more additional feeders along separate routes may be provided for critical loads, such as hospitals that cannot tolerate long interruptions. Switching from the normal feeder to an alternate feeder can be done manually or automatically with circuit breakers and electrical interlocks to prevent the connection of a good feeder to a faulted feeder.

Primary Network Systems

The primary network system consists of a grid of interconnected primary feeders supplied from a number of substations. It provides higher service reliability and quality than a radial or loop system. Only a few primary networks are in operation today. They are typically found in downtown areas of large cities with high load densities.

DISTRIBUTION TRANSFORMERS

Distribution transformers are of several types:

- Single phase or three phase;
- Pole mounted or pad mounted;
- Underground.

They come in various sizes usually small single phase units, and are filled with a dielectric fluid. The basic components of typical distribution transformers are shown in Figure 7.1. They can be purchased with various efficiencies, with better efficiencies costing more.

SECONDARY SYSTEMS

Secondary distribution delivers energy at customer utilization voltages from distribution transformers to meters at customers' premises. Figure 7.2 shows typical secondary voltages and applications in the United States. There are a number of types of secondary systems. Usually single phase, three-wire service is provided in residential areas. One of the three wires is a ground wire, the other two are energized. Connecting to the two energized wires provides 240 volts; connecting from either energized wire to the ground will supply 120 volts. Each transformer supplies a separate secondary system. In many

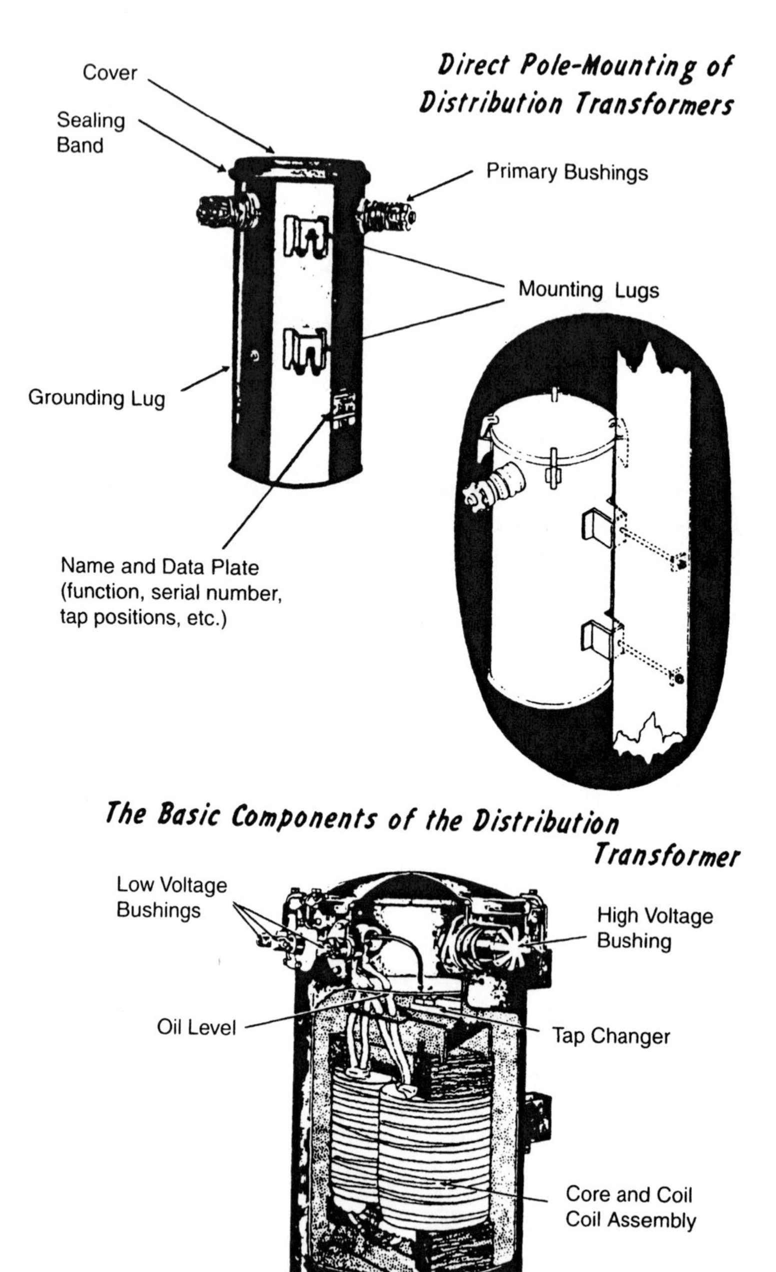

Figure 7.1. Typical distribution transformer.

Voltage	# Phases	# Wires	Application
120/240 V	Single phase	Three	Residential
208Y/120 V	Three phase	Four	Residential/Commercial
480 Y/ 277 V	Three phase	Four	Commercial/Industrial/High Rise

Figure 7.2. Typical secondary distribution voltages in the United States.

cases, there are tie-points between the secondary systems so that a supply may be obtained from an adjacent system if a transformer fails. Commercial and industrial loads are heavier than loads in residential areas, and a three-phase 4 wire supply is often installed since larger motors, used by these type customers, use three-phase power.

To supply high-density load areas in downtown sections of cities, where the highest degree of reliability is needed, secondary networks are used. Such networks are supplied by two or more primary feeders through network transformers. These transformers are protected by devices that open to disconnect the transformer from the network if the transformer or supply feeder are faulted. Special current limiting devices are also used at various locations in the secondary to keep problems from spreading. Smaller secondary networks called spot networks are also used to supply loads requiring extra reliability.

DISTRIBUTION CAPACITY

The capacity of the distribution system is determined in most cases by the thermal ratings of the equipment. In more rural areas with low load density it may be determined by voltage limits. The distribution substation capacity depends on the size of transformers and the provision of an additional spare transformer. If a substation has two transformers, all load must be supplied by the remaining one if one fails. In this case, the substation capacity will depend on the capability of the remaining transformer to carry the load for the time required to replace the failed transformer, with the capacity being lower if the replacement time is longer. For substations with a single transformer, load is limited to what can be transferred to other substations at remote feeder tie points.

The allowable primary feeder loading can be limited by the size of conductors used and the characteristics of the load supplied. If the load varies, higher maximum loads can be carried by the feeder than steady loads, since the rating of the feeder depends on the heating effect of the current over time. Feeder loading may also be limited by the voltage drop that occurs at the end of longer feeders.

The distribution transformer capacity is determined by the size of the transformer and the characteristics of the load. In some cases, the distribution transformers are single phase. When a three-phase supply is needed, three single-phase transformers can be used, each connected to a different phase of the three-phase primary supply. Alternatively, a three-phase transformer may be used in which the three phases are in a single tank.

The capacity of the secondaries is determined by the size of wire used, their length, and the characteristics of the load they supply.

LOSSES

Distribution systems have two types of energy losses: losses in the conductors and feeders due to the magnitude of the current and transformer core losses that are independent of current. Current related losses are equal to the current squared time the resistance of the feeder or transformer (I^2R). Accompanying these losses are reactive losses which are given by (I^2X). The core losses result from the energy used in transformer cores as a result of hysteresis and eddy currents. These losses depend on the magnetic material used in the core. As voltages vary from the design level, core losses can vary by as much as V^3 to V^5. Core losses in a power system can exceed 3% of the power generated constituting as much as 40% of the total loss on the system. The capacity of generation and reactive sources must be sufficient to supply these losses.

RATINGS

As in the case of transmission facilities, the ratings of distribution components are generally given as the product of the voltage at which they are operated and the current that is flowing through them (kVA). Normally seasonal ratings are used to recognize ambient temperature differences.

Distribution facilities are generally capable of operating at their rated value for specified periods of time for specific cycles, usually expressed as the "loss factor". The maximum rating and the period of time over which a component may be operated at its maximum rating depend upon the ambient temperature, the wind, sunlight, and the load levels experienced just prior to the time of the peak demand.

METERING

The maximum power and energy supplied the distribution substation, and the voltage in the substation, are continuously measured by recording meters in the substation. The energy used by each customer is continuously measured at the customer's meters. The electronic communication of customer meter readings to the utility for billing purposes and provision of data to assist operations and system design is increasing.

Significant changes in metering requirements have occurred as a result of the restructuring of the electric power industry to provide the ability to keep track of energy provided by various power suppliers. The use of telemetry is being increased to provide real time data. It helps reduce interruptions and accelerates restoration of power to customers.

CONTROL OF VOLTAGE

Good quality electric service requires that the voltage at the consumers' premises be kept within an acceptable voltage range for satisfactory operation of consumer equipment. At the 120-volt level, this is 110–126 volts at the utilization point. It is customary for utilities to hold voltage at the customer meter location between 114 and 126 volts, which allows for a 4-volt drop to the utilization point in the residence. The location of the voltage extremes are usually at the first and last customer locations on the primary feeder. During peak load conditions the first customer usually receives the highest voltage and the last customer the lowest. The variations from light to heavy load at these locations will establish the voltage range for the circuit.

As a first step in the control of voltage on such a circuit, most utilities will regulate the primary voltage at the substation. This takes care of variations in the voltage supplied to the substation and the variation on the feeder up to the first customers. The equipment usually used for regulation are tap changers on the substation transformers or separate feeder voltage regulators. For most urban feeders, no other regulating equipment is needed, although shunt capacitor banks are often installed to supply part of the kilovar portion of the load. On larger or longer feeders, both voltage regulators and shunt-capacitor banks may be needed out on the feeders to provide supplementary voltage control and reactive supply. In general, the control of voltage is more economical if both voltage regulators and shunt capacitors are used and if distribution voltage control is coordinated with voltage control of the transmission system and of generation.

Capacitors

Capacitors are applied as an economic tool to reduce system losses by supplying kilovars locally. Shunt capacitor banks including fixed and switched banks are used on primary feeders to reduce voltage drop, reduce power loss, and improve power factor. The closer to the load they can be installed, the greater the economic benefit. Capacitors are not only an economic tool for the distribution system, but they can eliminate the need for adding reactive sources in the bulk power system. Kilovars supplied directly to load areas reduce the current in all portions of the system. This releases transmission capacity and reduces system losses.

At light load, the capacitors installed for full load operation may cause too high a voltage on the distribution system. Therefore, many capacitors will have to be switched off during these periods. Various means are used to perform the switching.

Voltage Regulators

Voltage regulators usually are an autotransformer with automatic tap-changing under load. Automatic measuring and tap-changing equipment holds the output voltage within a predetermined bandwidth. By using the smallest practical bandwidth, more voltage drop can be allowed along the feeder, still keeping the consumer voltage within acceptable limits. The means for achieving this are an integral part of the regulator controls called the line drop compensator.

RELIABILITY

The distribution system is that portion of the electric power system which has the greatest direct impact on the level of reliability experienced by the consumer. Outage of a major generating unit might simply cause the electric utility to buy power from neighboring utilities or start up higher cost generating equipment available on their own system. Outage of a major transmission line might cause other transmission lines within the electric utility system to pick up additional load and require a redispatch of generation. However, outage of a single distribution feeder will usually result in immediate interruption of service to consumers directly connected to that feeder.

On overhead circuits, 80–90% of the faults are of a temporary nature caused by wind, lightening, icing, birds, small animals, and contact with tree limbs. If the fault is temporary, lasting only for a short time, the circuit can be successfully re-energized, restoring service to all consumers.

Protection of primary circuits against excessive currents is provided by circuit breakers, automatic circuit reclosers, fuses, and sectionalizers, which divide the primary circuit into a number of sections. The time-current characteristics and operating characteristics of these devices are coordinated so that service is restored to all consumers following a temporary fault, and a minimum number of consumers are interrupted for a permanent fault.

Reclosing circuit breakers and automatic circuit reclosers have both an instantaneous overcurrent characteristics and a time-delay overcurrent characteristic. Initially, these devices trip instantaneously, interrupting the fault current quickly enough to prevent the blowing or melting of downstream fuses. If the fault persists when the circuit is reclosed, these devices switch to a time-delay trip characteristic. This permits downstream fuses to blow and isolate a permanent fault before the breaker or recloser trips. Automatic reclosing is generally not used on cable circuits to prevent increasing damage from a cable fault.

Standards have been established for measuring and comparing the reliability provided to distribution customers. These are given in Chapter 9.

Quality of Service

Distribution systems are also subject to voltage dips and other variations in quality of service. These can be caused by the effects of other customer's apparatus or by faults or short-circuits at other points on the system. When a fault occurs, the voltage at the fault location will go to zero. Voltages at nearly locations will be significantly depressed for the duration of the fault. Voltages will be depressed to lesser degree at more remote locations. But transmission system faults can cause dips at locations as far away as 100 miles, depending on the voltage level of the faults. These voltage dips affect digital clocks, computers, and other electronic devices and has lead the industry to address improvements in the reliability and quality of service.[2]

DESIGN OF DISTRIBUTION SYSTEMS

A reliable distribution system must be designed to meet future power supply requirements. It must also have adequate protection for the various types of faults and short-circuits that can occur. This requires that circuit breakers, fuses, and other protective devices have the ability to interrupt the very high currents that can occur when a short-circuit occurs. Relaying to detect such faults needs to be provided and coordinated with the protective devices.

There is a considerable amount of software available for the design and operation of distribution systems, including:

- Capacitor placement optimization;
- Circuit breaker duties;
- Conductor and conduit sizing—ampacity and temperature computations;
- Database management;
- Distribution reliability evaluation;
- Distribution short-circuit computations;
- Graphics for single-line diagrams and mapping systems;
- Harmonics analysis;
- Motor starting;
- Power factor correction;
- Power flow/voltage drop computations;
- Power loss computations and costs of losses.

[2] EPRI, The *Cost of Power Disturbances for Industrial and Economy Companies*, 2001.

DISTRIBUTED GENERATION

The increasing application of small generation sources on the distribution system is being driven by economics, reliability concerns and the development or new technology. Some distributed generators will be installed by the utility on the supply side of the customers' meters. Others will be installed by the customers on their side of the meter and will affect the distribution system and the other customers it supplies. The connection of diesels, fuel cells, photo-voltaic cells, wind generators and micro turbines raises new concerns and problems for distribution systems. They increase short-circuit duties, they all require relaying and protection changes, and raise questions of safety both to utility workers and consumers.

The use of distributed generation to help provide power impacts the design and operation of bulk supply systems, including ancillary services such as reactive capacity, spinning reserve, and so on. Coordination is needed in the design and operation of the bulk supply system and the distribution system. This is complicated by the fact the bulk power and the distribution systems generally are owned by different parties and separately regulated with the Federal Government regulating the bulk supply and the State Government regulating the distribution systems. One alternative is the development of coordination contracts between the parties which will provide for an equitable sharing of the benefits from coordination.

OPERATION OF DISTRIBUTION SYSTEMS

The distribution system is operated and controlled using SCADA (system control and distribution automation) system at a dispatch center. These systems are of various types and are under continuous development. They are used to provide data needed for operation and new billing requirements.

The automation of the distribution systems continues to increase. The benefits of distribution automation include:

- Improved distribution reliability;
- Reduced customer outages and outage durations by automatically locating and isolating faulted sections of distribution circuits and automatically restoring service to unfaulted sections;
- Reduced customer complaints;
- Reduced power losses for substation transformers, distribution feeders, and distribution transformers;
- More effective use of distribution through automatic voltage control, load management, load shedding, and other automatic control functions;
- Improved methods for logging, storing, and displaying distribution data;

- Improved engineering, planning, operating, and maintenance of distribution.

The current status of distribution automation development is shown in Figure 7.3.

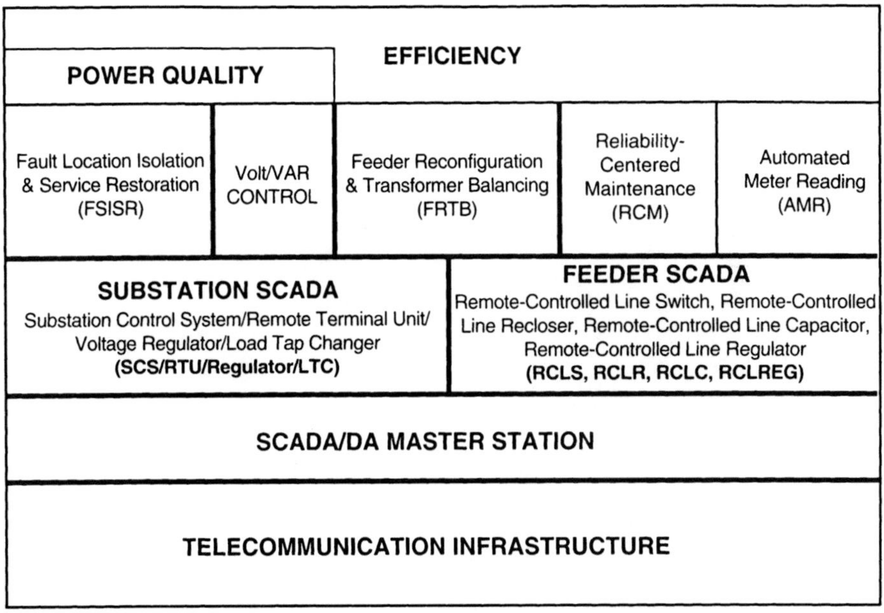

POWER QUALITY

EFFICIENCY

Fault Location Isolation & Service Restoration (FSISR)

Volt/VAR CONTROL

Feeder Reconfiguration & Transformer Balancing (FRTB)

Reliability-Centered Maintenance (RCM)

Automated Meter Reading (AMR)

SUBSTATION SCADA
Substation Control System/Remote Terminal Unit/ Voltage Regulator/Load Tap Changer
(SCS/RTU/Regulator/LTC)

FEEDER SCADA
Remote-Controlled Line Switch, Remote-Controlled Line Recloser, Remote-Controlled Line Capacitor, Remote-Controlled Line Regulator
(RCLS, RCLR, RCLC, RCLREG)

SCADA/DA MASTER STATION

TELECOMMUNICATION INFRASTRUCTURE

Figure 7.3. Automation of electric distribution systems.[‡‡‡‡]

[‡‡‡‡] Extracted from a DISTRIBUTION AUTOMATION: A STRATEGIC OPTION FOR ELECTRIC UTILITIES IN THE RESTRUCTURED AND DEREGULATED ENVIRONMENT—Manuel C. Mendiola, Member, IEEE Manila Electric Company.

8

FUNCTIONING OF THE ELECTRIC BULK POWER SYSTEM

So far we have discussed the elements of the electric system. This chapter addresses how these elements come together to ensure that electricity is available when needed. The chapter will cover both the operation and the planning of the system.

The process by which the electric bulk power system functions has both a technical and organizational dimension. At the time of writing this book, the organization structures are undergoing change due to restructuring efforts under way at the national and state levels.

COORDINATION

The operation of the bulk power system in the United States reflects the interdependency of the various entities involved in supplying electricity to the ultimate consumers. These interdependencies have evolved as the utility industry grew and expanded over the century.

Since the electric system operates in large synchronous interconnections, the effects of power flows and electrical disturbances are felt by all systems connected to a synchronous grid. In a power system, the coordination of all elements of the system and all participants are required from an economic and a reliability perspective. Generation, transmission, and distribution facilities must function as a coordinated whole. The scheduling of generation must recognize the capability of the transmission system. Voltage control must involve

Understanding Electric Power Systems: An Overview of the Technology and the Marketplace, by Jack Casazza and Frank Delea
ISBN 0-471-44652-1

generators as well as transmission and distribution facilities. Actions and decisions by one participant, including decisions not to act, affect all participants.

In parallel with its early growth, the industry recognized that it was essential that operations and planning of the system be coordinated and organizations were formed to facilitate the joint operation and planning of the nation's electric grid. Initially holding companies and then power pools were established to coordinate the operation of groups of companies.

After the Northeast Blackout of 1965, regional electric reliability councils were formed to promote the reliability and efficiency of the interconnected power systems within their geographic areas. These regional councils joined together shortly afterwards to form a national umbrella group, NERC—the North American Electricity Reliability Council. At present there are ten regional councils. The members of these Regional Councils come from all segments of the electric industry: investor-owned utilities; federal power agencies; rural electric cooperatives; state, municipal and provincial utilities; independent power producers; power marketers; and end-use customers. These entitles account for virtually all the electricity supplied in the United States, Canada, and a portion of Baja California Norte, Mexico.[1]

Since its formation in 1968, the North American Electric Reliability Council (NERC) has operated as a voluntary organization to promote bulk electric system reliability and security—one dependent on reciprocity, peer pressure, and the mutual self-interest of all those involved.

In promoting electric system reliability and security, NERC, among other things:

- Establishes operating policies and planning standards to ensure electric system reliability;
- Reviews the reliability of existing and planned generation and transmission systems;
- Critiques past electric system disturbances for lessons learned and monitors the present for compliance and conformance to its policies;
- Maintains liaisons with the federal, state, and provincial governments in the United States and Canada and electricity supply industry organizations in both countries.

The growth of competition and the structural change taking place in the industry have significantly altered the incentives and responsibilities of market participants to the point that a system of voluntary compliance is no longer adequate. In response to these changes, NERC is in the process of transforming itself into an industry-led self-regulatory reliability organization (SRO) that will develop reliability standards for the North American bulk electric system.

[1] Extracted from About NERC, www.NERC.com.

After FERC issued its Orders 888 and 889 in 1996, some areas of the country formed Independent System Operators (ISOs). The New York and PJM ISOs mission statements are typical:

> New York—"... to ensure the reliable, safe, and efficient operation of the State's major transmission system and to administer an open, competitive, and non-discriminatory wholesale market for electricity in New York State."
>
> PJM—"Maintain the safety, adequacy, reliability, and security of the power system. Create and operate a robust, competitive, and non-discriminatory electric power market."

Recently, in response to FERC directives, Regional Transmission Organizations (RTOs) are being formed. These organizations will be responsible for the planning and operation of their respective bulk power grids.

OPERATION

Control Areas

The overriding objectives of those individuals responsible for the performance of the electric system is to ensure that at every moment of time there is sufficient generation to reliably supply the customer requirements and all associated delivery system losses. The process is complicated by the fact that the customer load changes continuously and, therefore, the generation must adjust immediately, either up or down, to accommodate the load change. Since electric power cannot be stored, the generation change must be accomplished by a physical adjustment of the equipment generating the electricity.

To coordinate the operation of the bulk power system, a system of control areas has evolved. Each control area is responsible for maintaining its own load/generation balance including its scheduled interchange, either purchases or sales. A control area can consist of a generator or group of generators, an individual company, or a portion of a company or a group of companies providing it meets certain certification criteria specified by NERC. It may be a specific geographic area with set boundaries or it may be scattered generation and load.

Figure 8.1 shows the location of each of the reliability councils as well as the location of the more than 140 control areas within NERC. The control areas vary greatly in both geographic size and the amount of generation/load they control. One of the results of the on-going process to set up RTOs will be a reduction in the number of these areas.

Each control center maintains communication with adjoining areas in order to coordinate operations. Coordination activities can include inter-area power exchanges, power transfer limits on interties and circulating or inadvertent power flows.

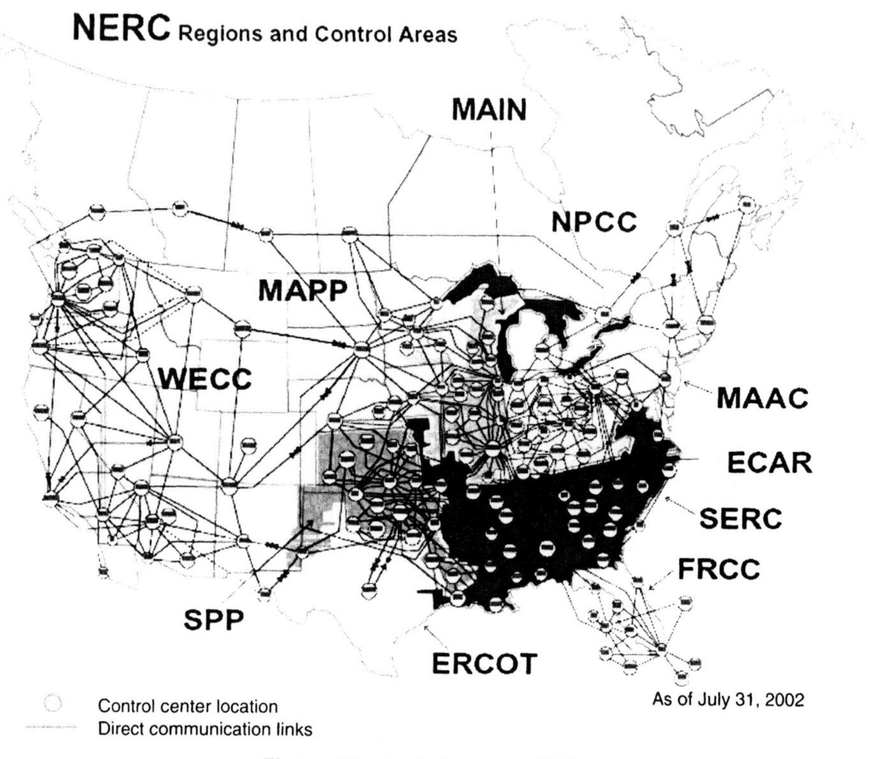

Figure 8.1. Control areas in NERC.

In addition to maintaining communication with other control areas, the control center has to be in contact with its constituents, be they companies or individual generating stations or generating units or individual substations. If they are in contact with an individual company, that company, in turn, has communication links with its generating units or transmission operating centers so that control signals can be passed rapidly.

Since minute-by-minute customer load change are not known in advance, a system has been developed where generation changes are made in response to load changes. This system is based on the concept of the area control error.

The control centers also require real time information about the status of the system. This information includes power line flows, substation voltages, the output of all generators, status of all transmission lines and substation breakers (in-service or out-of-service) and transformer tap settings. Some areas are implementing real-time transmission line rating systems requiring additional information such as weather conditions, conductor temperatures, and so forth.

Each control area monitors on an on-going basis the power flow on each of its interties (in some cases delivery points) and the output of each generator within its control. The sum of the internal generation and the net flow on the interties is equal to the customer load and all transmission losses within the area.

The various commercial interests that are involved within the area are required to notify the control area personnel of their contractual arrangements on an on-going basis for either sales or purchases of electricity with entities outside the area's boundaries. Additionally, neighboring operating entities, engaged in transactions that will cause power to flow through the control area are required to notify the control area and to make provision for the attendant transmission losses.

The reporting requirements have been formalized by FERC in its Orders 888 and 889, issues in 1996. The Order stated, in part, that:

> "A public utility . . . must rely on the same electronic information network that its transmission customers rely on to obtain information about its transmission system when buying or selling power."

The reports are entered into the Open Access Same-Time Information System (OASIS), an Internet based bulletin board which gives energy marketers, utilities, and other wholesale energy customers, real-time access to information regarding the availability of transmission line capability. OASIS provides the ability to schedule firm and non-firm transactions.

With this information, the control area operators can compare the total scheduled interchange into or out of the control area with the actual interchange. If the receipt of electricity exceeds the schedule, the control area must cause the scheduled generation levels to increase. If the receipt is too low, scheduled generation within the control area is reduced. These adjustments are typically made a day ahead and then in real time. Since these adjustments are on-going simultaneously by all control areas, the adjustments balance out.

The process where individual contracts scheduled within OASIS are identified as to source and customer is known as tagging. This information, while it may be commercially sensitive, is critical if system operators are to adjust system power flows to maintain reliable levels.

Concurrently, the system operators can also evaluate the expected power flows internal to the control area to determine if adjustments are required in the generation pattern to insure that all transmission facilities are operated within their capabilities.

Each control area also participates in maintaining the average system frequency at 60 hertz. The system frequency can deviate from normal when a large generating unit or block of load is lost.[2] In addition to adjustments made because of variations of tie flows from schedule, another adjustment is made to correct frequency deviations.[3] Each control area is required to have an

[2] Generators larger than 10 mW within each control area should be equipped with speed governors which will respond to frequency (i.e., speed) excursions.

[3] This is called automatic generation control (AGC).

adjustment factor related to frequency in its control logic. The term is called the tie-line frequency bias (expressed in mW/0.1 Hz).

Additionally, since the control process is responsive, there can be a drift in average system frequency, which, in turn, affects the accuracy of any electric clocks. This variation is monitored and for a period of time the target frequency reference is adjusted to produce the required compensation. This process is called Time Error Correction.

There are a variety of market rules, which vary across the country, by which the output of individual generators is scheduled. These rules are currently in a state of flux as the FERC moves to a free market system for dispatching generation. We will discuss issues attendant to this effort later in the book.

With the restructuring of the industry, the emergence of merchant power plant owners, the development of ISOs, RTOs, and for-profit transmission companies, and the implementation of retail access in some regulatory jurisdictions, assigning all reliability responsibilities to control areas makes the job of defining and applying Standards more and more complicated. This is further complicated since some control areas are acting as transmission service providers. NERC has an effort under way to rationalize this process. It's begun by defining almost 100 reliability functions that need to be addressed.

Operating Reserves

Each control area must provide operating reserves to restore its tie flows to schedule within 15 minutes following its most severe single contingency. Operating reserves consist of spinning and non-spinning reserve. Spinning reserve is generation that is synchronized and available to supply incremental load in a specified time period. Non-spinning reserve is not synchronized but can be made available within a short period of time. Interruptible load disconnection and coordinated adjustments to interchange schedules can be considered as part of operating reserve.

Each control area must supply reactive resources within its boundaries to protect its voltage levels under contingency conditions.

While balancing load and generation, each control area is responsible for ensuring that the bulk power system is operated to meet a consistent set of criteria, standards and procedures. NERC has established operating policies and planning standards to ensure electric system reliability. The specific details have been codified by each of the ten regional reliability councils. In addition, local areas may implement more stringent criteria, standards and procedures, if their situation warrants.

Control areas are also required to coordinate maintenance and protective relaying and to have a system restoration plan.

Ancillary Services

FERC in its Order 888 specified Ancillary Services needed to facilitate the operation of a bulk supply system. These services were grouped into three broad categories. They are:

- Required to be provided by transmission providers:
  - Scheduling, system control and dispatch;
  - Reactive supply and voltage control by generators;
- Required to be offered by transmission provider;
  - Regulation (AGC);
  - Operating reserve—spinning;
  - Operating reserve—supplemental;
  - Energy imbalance (hourly mismatches).
- Not required of transmission providers:
  - Load following (hourly or daily);
  - Back-up supply;
  - Real power loss replacement;
  - Dynamic scheduling;
  - System black start capability;
  - Network stability services.

Issues attendant to FERC's Order 888 will be covered more fully later in this book.

Emergencies

At present, two basic philosophies exist concerning potential emergencies.

1. Preventative Philosophy. One approach that has predominated in the United States for many years is preventative operation. What this means is if a system operator or a pool operator discovers a condition on his system where a single contingency will cause an overload and possible trip out of another facility, he will adjust the operation of his system to reduce the loading conditions to eliminate this potential hazard. With this philosophy power transfers, interchanges, and economic dispatches which could be economically beneficial are often not made. This is the philosophy recommended by the North American Reliability Council.

2. Corrective Philosophy. There is an increasing amount of attention being given to the impact of taking larger risks in the operation of transmission systems because of the significant savings that might be made. The corrective philosophy provides that you do not reduce the transfers until after the contingency has occurred. This means that for the high percentage of the times that the contingency does not occur, the economic benefits will accrue. However, it does mean for the small percentage of the time when the contingency does occur that severe reliability penalties could result through a major disruption of the power supply.

The preventative approach is generally used throughout the United States. Eventually both the capacity and reliability of transmission networks will have to be improved simultaneously through development of a highly automated, "smart" power system.[4] The grid will need technological advances in four major areas:

1. Improved physical control to expedite grid operations by switching power more quickly and preventing the propagation of disturbances;
2. Monitoring systems that can improve reliability by surveying network conditions over a wide area;
3. Analytical capability to interpret the data provided by the wide area monitoring system for use in network control; and
4. A hierarchical control scheme that will integrate all the above technologies and facilitate flexible network operations on a continental scale.

Electric systems are now adding these technologies to their transmission systems, creating smart networks. The possible future of these "smart" control schemes will have to be carefully analyzed to recognize and evaluate such a potential future.

Operating Emergencies

NERC requires each system, control area, pool and Region to have a set of plans to cope with operating emergencies. These plans must cover:

- Agreements for emergency assistance from neighboring systems;
- Procedures for system operators;
- Authority to shed load to restore generation/load balance;
- Procedures for system restoration.

Items to be considered include fuel supply and inventory, fuel switching at generators where possible, environmental constraints, reduction of in-system energy use, public appeals, implementation of load management and voltage reductions,[5] appeals to large customers to reduce usage, running generators at maximum output, requests for governmental aid to implement energy reductions, and mandatory involuntary load curtailment.

In the event the above actions are insufficient or if the situation develops too rapidly to implement them, each area is required to install underfrequency load shedding. This is a system where protective relays detect a low-frequency condition and actuate the disconnection of blocks of load in an attempt to arrest the decline in frequency and restore 60 Hz operation before the low-frequency results in the loss of additional generation.

[4] K. Stahlkoph & P.R. Sharp. 2000. *Where Technology and Politics Meet, Electric Power Transmission Under Deregulation*. IEEE.

[5] The effects and duration of load reductions depend on the characteristics of the load.

PARALLEL PATH FLOW AND LOOP FLOW

Parallel power flows reflect the interconnected nature of the bulk power system that we have mentioned previously. Since power flows on all transmission paths, it is not uncommon to find circumstances where part of a power delivery within one control area flows on the transmission lines in adjoining areas or where a part of a power delivery between two control areas flows over the transmission facilities of a third area. In addition, loop flows also occur and are the result of generation location and transmission system design and reflect the situation where all systems are supplying their own load from their own resources but result in power flow on the transmission in other systems. These circumstances have been referred to as a seams issue.

POWER TRANSFER LIMITS

A primary aspect of a control center's responsibility toward the reliability of the bulk power system is to make certain that the levels of power transfers that take place are within the capability of the bulk power transmission system reflecting that area's operating criteria.

In order to define the amount of transmission capacity available for commercial transactions, NERC[6] developed the following definitions:

$$\begin{aligned} &\mathit{Available\ Transfer\ Capability\ (ATC) = Total\ Transfer\ Capability\ (TTC)} \\ &\quad - \mathit{Existing\ Commitments} - \mathit{Transmission\ Reliability}\ \text{Margin}\ (\mathit{TRM}) \\ &\quad - \mathit{Capacity\ Benefit}\ \text{Margin}\ (\mathit{CBM}), \end{aligned}$$

where:

- Available Transfer Capability (ATC) is a measure of the transfer capability remaining in the physical transmission network for further commercial activity over and above already committed uses.
- Total Transfer Capability (TTC) is the amount of electric power that can be transferred over the interconnected transmission network in a reliable manner while meeting all of a specific set of defined pre- and post-contingency system conditions.
- Transmission Reliability Margin (TRM) is the amount of transmission transfer capability necessary to ensure that the interconnected transmission network is secure under a reasonable range of uncertainties in system conditions.

[6] "Available Transfer Capability Definitions and Determination"—NERC, June 1996.

- Capacity Benefit Margin (CBM) is that amount of transmission transfer capability reserved by load serving entities to ensure access to generation from interconnected systems to meet generation requirements in emergencies.

Determination of Total Transfer Capability

The total transfer capability is the acceptable magnitude of power flow in one direction over a group of transmission lines often referred to as an interface or a flowgate. A control area may have many of these interfaces both internal to the area and at its points of connection with adjoining areas. The transmission lines comprising an interface are not necessarily at the same voltage or between the same substations.

The capability is determined by examining the performance of the transmission system for a variety of contingencies to determine if the contingency would occur, the level of power flow when:

- The loading of an individual transmission line exceeds that line's capability;
- The bulk power voltage at any substation would be unacceptably low;
- The system would become unstable if a contingency occurred.

The Total Transfer Capability is set by the power flow level at the most limiting constraint. It can be:

- Pre-disturbance related:
  - Unacceptable line loadings or bus voltages.
- Disturbance related:
  - Transient or dynamic instability.
- Post-disturbance related:
  - Unacceptable line loadings or bus voltages.

For a stability limit or a voltage limit, none of the transmission lines on the interface may be loaded to its individual capacity. For a situation where the loading on an individual line sets the limit, the loadings on all parallel lines are most probably below (perhaps well below) their capacities. It is these latter situations that are the reason for interest in the use of phase-shifting transformers and in the development of FACTS devices, which would allow the system to transmit more power with existing facilities.

The contingencies to be considered when determining the Total Transfer Capabilities are detailed in NERC and individual reliability council criteria. As a minimum, the bulk power system is to be operated to what is called the N – 1 criteria. NERC states, "All Control Areas shall operate so that instability, uncontrolled separation, or cascading outages will not occur as a result of the most severe single contingency."

Reduction of Power Transfers—Congestion Management

Congestion is a term that is applied to situations where the amount of power flowing or projected to flow across a group of transmission lines (a flowgate or interface) exceeds its capacity. To relieve the potential overload, the electrical operation of the system must be adjusted:

- Reducing power flows by reducing generation in the sending system and increasing generation or reducing load in the receiving system, or
- Changing the system configuration by opening lines or by closing or segmenting busses.

This process is covered by NERC's Transmission Loading Relief Procedure.

With the on-going efforts to open the electricity systems for market based wholesale transactions, complications arise when attempting to redispatch the system to eliminate or avoid power transfer limit violations—specifically, how should economics be factored into the adjustments? One complication is that the generating units that might have the most direct effect on reducing the transfer limit violation might be lower cost than other units further removed from the violation. A further complication occurs when the various generators that could relieve the violation are in areas using different market rules. This is another aspect of the seams issue.

Congestion limits are essentially economic limits concerned with operation of the power market. The location of these limits is not always the same as the location of the actual reliability risks in the transmission system (this is discussed further in the reliability chapter).

PLANNING

Prior to the opening of the transmission system to many different users, the planning process was reasonably straightforward. The process integrated load forecasting, generation planning and transmission planning and was usually done by one entity, either a company or a power pool. Coordination of plans was done under the auspices of the Regional Councils.

The process normally started with a peak load forecast. In most cases the forecast projected a growth in peak loads.[7] The planning objectives were:

- Generation—to have enough generation capacity to meet the projected peak load plus a reserve margin;
- Transmission—to connect generators to the grid, to have enough transmission capability to reliably deliver generation and firm purchases to

[7] Many utilities experienced financial difficulties during the 1970s and 1980s when forecast peak loads did not materialize, while, at the same time, costly, large generating units were under construction.

existing and new load centers, to accommodate the sharing of reserves with nearby areas, and to allow economically driven power exchanges both intra-area and inter-area;

- To provide these services over an extended period of time at minimum cost.

In both cases, the financial return on and of the resulting facilities was regulated and based on a cost-of-service perspective.

Regional Councils had agreed upon criteria covering:

- Generation reliability expressed in terms of a target reserve or minimum acceptable loss of load probability; and
- Transmission reliability expressed in terms of a number of disturbances that the transmission system had to be able to withstand while meeting the above stated objectives.

PLANNING STANDARDS

NERC's Planning Standards[8] define the reliability aspect of the interconnected bulk electric systems in two dimensions:

1. Adequacy—the ability of the electric systems to supply the aggregate electrical demand and energy requirements of their customers at all times, taking into account scheduled and reasonably expected unscheduled outages of system elements; and
2. Security—the ability of the electric systems to withstand sudden disturbances such as electric short circuits or unanticipated loss of system elements.

To these we should add safety for workers and the general public and, especially since September 11th, 2001, an expanded sense of the meaning of the term security.

Generation Planning

Generation reserve margin targets were established using probability techniques relating generating availability and potential load forecast errors to a probability that for some hours of the year the load could not be supplied. These reserve margins were formally established as design requirements. The most commonly used target was one day in ten years.

The determination usually reflected, for each year, statistics on the relia-

[8] NERC Planning Standards at www.NERC.com.

bility of individual generators, the expectation of hourly peak loads, the effect of aid from nearby systems, intra-area transmission capabilities and various levels of remedial actions by operators.

These probability techniques modeled the effects of having interties with neighboring systems and they demonstrated that relying on emergency assistance from neighboring areas could result in dramatic reductions in installed reserve requirements for both areas, providing that the transmission interties were capable of supporting the power transfers. The probabilistic techniques are called loss of load probability for capacity constrained systems, or loss of energy probability for energy constrained systems.

Recently, the need for targeted reserve margins has been questioned. The issue is whether the marketplace should determine the level of installed reserve. Some areas require that load servicing entities ensure that they have sufficient capacity to supply their peak load and to have a reserve margin available.

Once the amount of load was determined, the generating capacity required to achieve the targeted reserve levels was known. The next step was to decide on the specific generating types that would comprise the mix. The starting point was always the existing installed base of generation. Analyses were made to determine whether any of these generators should be retired during the forecast period. Typically, generator units might have useful lives of 40 or more years. In a regulated system, these older generators might have significantly higher operating costs than newer generators but these were somewhat offset by a reduction in their capital investment due to depreciation. The resulting capital and operating costs were rolled into an average system rate charged to customers. For some existing units an option was a major overhaul and rebuilding sometimes referred to as life extension.

New-generation options had two dimensions; the technology to use and whether the utility should go it alone or become part of a group building a unit to share the financial burden and the risks. Generation types were selected based on the number of hours the new unit was expected to run. For example, if the capacity needed was only during peak hours, a peaking unit would be selected. Peaking units would involve a low capital cost, relatively high operating costs and with short lead times. Conversely, if the unit were to run almost continuously, a base load design would be selected. These units typically had high capital coats but low operating costs. Availability of sites for the generation including land, cooling water, and means of delivering fuel were considered as well as the costs and availability of various fuel options. Expansion plans covered multi-year periods so that individual decisions to add specific types of capacity were made in a broader context.

The options were screened and evaluated using programs which modeled the yearly dispatch of all units in the system or power pool in which the unit would operate. These production costing programs considered yearly load shapes, generation maintenance and unavailability, sales and purchases from adjoining areas, intra and inter-area transmission capacity, and individual generator fixed and variable running costs including fuel costs versus electrical outlet (heat rates). Economy sales usually were made based on splitting the resulting savings equally.

Least Cost Planning

Past planners evaluated the means by which the forecasted peak load could be supplied. Starting in the 1970s, a process known as least cost planning, came into wide use. Trade-offs were made between adding new generation and instituting programs to reduce customer peak load.[9] As discussed in the customer load section, peak loads occur for very few hours a year. Reducing the peak load in some cases was a less costly alternate than building new generation. At the time, the financial yardstick was not profits but rather minimum revenue requirements from customers.

Transmission Planning

NERC specifies transmission systems planning standards that cover the types of contingencies that must be examined for conditions for all facilities in service and with facilities out-of-service for maintenance while delivering generator output to projected customer demands and providing contracted firm (non-recallable reserved) transmission services, at all demand levels. These contingencies can result in the loss of single or multiple components. For each of the contingencies, the system must be stable and applicable thermal and voltage limits must be observed. For the loss of multiple components, the controlled interruption of customer demand, the planned removal of generators, or the curtailment of firm (non-recallable reserved) power transfers may be necessary.[10]

The standards also require evaluation of the risks and consequences of a number of extreme contingencies such as the loss of all circuits on a R-O-W, all generators at a generating station, or failure of circuit breakers to clear a fault. Individual Regions may develop their own regional planning criteria to reflect circumstances applicable to their own situation. These Regional Criteria are evaluated by NERC to ensure consistency with NERC's planning standard.

NERC also covers in its Planning Standards:

- Reliability assessment;
- Facility connection requirements;
- Voltage support and reactive power;
- Transfer capability;
- Disturbance monitoring.

[9] Over time pressure mounted in some jurisdictions to expand the customer programs from a peak load reduction focus to an energy reduction focus.

[10] The Table summarizing the Standards has not been included in this document. The reader can access the table at www.nerc.com, Planning Standards.

The present complexity of the NERC Standards reflects the changing state of the electric utility industry. When NERC and the Regional Councils were first formed, their membership was almost entirely utilities and the rules, standards, best practices that were produced relied on voluntary observance by the members. As the industry has moved to its present structure with many more participants, NERC has been working to make the planning and operating rules for the utility industry clear, universal, and well documented. To do this has meant that the volume of the associated material has grown to a point that no one text could hope to cover it all. As the footnotes indicate, individuals wishing more detail on these matters can find them at NERC's Web site: www.nerc.com.

As a generation expansion pattern was being developed, transmission planners would address the transmission expansion needed to accommodate the generation and the forecast load growth. Development of a transmission plan has been described as part science and part art. There are three situations that confront the transmission planner:

1. Connect a new generator or generating station to the grid.
2. Connect a new substation to the grid.
3. Reinforce the existing grid.

The obvious first step for connecting a new generator or a new distribution substation is to build one or more lines to the nearest bulk power substation. However, this may not be sufficient or adequate. An examination is needed to see if the capability of the existing grid is sufficient to accommodate either. This examination has to consider a wide range of operating conditions including different load levels, different power transfer patterns on the grid, and various maintenance outages. The analysis should evaluate a number of years into the future including additional generation and distribution substation requirements. It may well be that because of future developments, larger more robust facilities should be installed initially or, even, that future expansion may mitigate the need for facilities now. For example, if generation is presently being sited outside a generation deficient load area, an initial reaction might be to build a large-scale transmission development into that area. What if, however, subsequent generation additions are within the generation deficient area? The result could be that the transmission additions could be lightly loaded and do not carry enough power to pay for their costs.

An important consideration is that the transmission additions may not always be near the new generation. The restriction on the capacity of a section of the grid can be far removed from the new generator addition. Other instances have been seen where new lines are added to increase stability margins although they carry little if any power themselves.

After examining the need over a sufficiently long time span, decisions are needed on the voltage level of the new line(s), their thermal capacity, their terminal locations, and the circuit breaker arrangements at these locations.

Load-Flow Studies. The transmission planning process uses a simulation program called a load-flow or a power-flow. This program solves Kirchoff's equations for a moment in time. It provides, for each condition for a given network topology, load level and generation schedule, the resultant real and reactive power flows in each line and transformer, the voltage at each bus and the mVAr output of each generator.

The simulation requires an enormous amount of information from the individual utility and from all utilities whose operations might interact with the planner's system. Due to the interconnected nature of the bulk power transmission system, this requires modeling of large geographic sections of the country in enough detail to capture the effects of power flows on all adjacent systems or resulting from the generation/load patterns in all adjacent systems. The model allows each control area to be modeled separately. Each control area is identified usually by attaching an identifier to each bus and a targeted interchange is specified. The mW output of one generator in each area is designated as being variable to balance the area's interchange to the desired level.

Repeated simulations allow the planner to evaluate the performance of the system at various levels of power transfers before and after various contingencies such as the loss of a transmission line, transformer, or generator. In some areas of the country since the electric system is considered electrically tight,[11] the planner can interpolate the results of various contingencies to determine the limiting contingency, the limiting element, and the limiting level of power flow.

Stability Studies. Another simulation program used in the transmission planning process is called a stability program. It uses, as a starting point, a solved load-flow program. It expands the representation of generation to include various levels of detail about the electrical and mechanical characteristics of the turbine generator, including generator reactance, excitation system characteristics, and turbine characteristics. Representations of the protective relay characteristics can also be included.

The program simulates the performance of an electric system over a period of time, usually a few seconds, following a disturbance. The length of time of the simulation and the complexity of the electrical effects being modeled determines the amount of detail provided about the turbine generators.

As noted in the discussion of load-flow studies, obtaining the required information for these studies is a major endeavor.

Short-Circuit Duty Studies. In some areas short-circuit duties are a problem. Short-circuit duties refer to the ability of a circuit breaker to inter-

[11] The electrical angle across the system is small such that variations in the electrical performance can be considered to be linear.

rupt the current that flows to a fault or the ability of all the elements of the substation (bus, transformers, etc.) to physically withstand the mechanical forces that result from the fault current. Considerations involved are: the number of phases involved in the fault, the impedance of the fault, and the electrical proximity of the generating sources (sources of current).

Planners must evaluate the interrupting capability required at all new substations and at all existing substations since both will be affected by new generation. In most instances, the interrupting capability can be achieved by purchasing circuit breakers of sufficient capacity and/or by upgrading existing breakers. In some instances, the required circuit breaker capacity cannot be purchased. In other cases a complete substation rebuild may be required to provide needed mechanical strength at a very high cost. Other solutions may be needed. These solutions include:

- Install fault current limiting devices such as series reactors;
- Changing the system configuration by opening bus ties or using back-to-back dc.

NEW PLANNING ENVIRONMENT

Both the objectives and the process of planning have changed. The degree of change depends on the area of the country and how far local regulatory authorities have moved the restructuring process, that is, has generation divestiture and/or retail access been implemented? As a matter of national policy, wholesale power producers are encouraged and given access to the transmission grid under the same terms and conditions as the local utilities.

The generation planning process has changed in the new environment to the extent that a global approach is no longer used. NERC no longer has criteria defining the minimum acceptable level of generation reliability. Some regional areas retain a required generation reserve, some do not.[12] Generation pricing is based on the market not on costs. With many different companies building power plants, each decision to build is made with less certainty of what the competitive situation will be going forward. The selection of unit sizes and types is made with a view towards those that will be the most profitable, not necessarily towards those that will result in the overall lowest costs to the consumer. Many felt, perhaps foolishly, that the market would drive prices down.

The design objective for the transmission system is being expanded to include provision for sufficient transmission capability to facilitate a geographically wide scale wholesale power market. FERC is actively seeking change in the way the transmission system is planned:

[12] FERC, in its proposed standard market design, specifies a level of generation reserve of at least 12%.

- In its Order 2000, it specifies one of the minimum functions of an RTO as, in part, that it "... must develop a planning and expansion proposal that (1) *encourage market-motivated operating and investment actions* (emphasis added) for preventing and relieving congestion."
- One of the objectives in its Strategic Plan for FY 2002–2007 is to "Firmly establish transmission planning function on a regular basis, with a variety of technology solutions to meet reliability, security, and market needs."

It is fairly typical to have a lengthy list of proposed new generation in each state. However, only some of the proposals are being implemented. Approval of new power plants is done under regulatory rules which vary state by state. In some states the process is relatively rapid, in others it is not. Merchant power plant owners obviously will opt for construction in supportive regulatory jurisdictions. This then translates into an increased need for transmission capability to move power from these states to states where power plant construction is hindered, since these latter markets probably will have higher market clearing costs.

NERC also includes the notion of economic interchange when it defines the purposes of the transmission system:[13]

- "Deliver Electric Power to Areas of Customer Demand—Transmission systems provide for the integration of electric generation resources and electric system facilities to ensure the reliable delivery of electric power to continuously changing customer demands under a wide variety of system operating conditions.
- Provide Flexibility for Changing System Conditions—Transmission capacity must be available on the interconnected transmission systems to provide flexibility to handle the shift in facility loadings caused by the maintenance of generation and transmission equipment, the forced outages of such equipment, and a wide range of other system variable conditions, such as construction delays, higher than expected customer demands, and generating unit fuel shortages.
- Reduce Installed Generating Capacity—Transmission interconnections with neighboring electric systems allow for the sharing of generating capacity through diversity in customer demands and generator availability, thereby reducing investment in generation facilities.
- Allow Economic Exchange of Electric Power Among Systems—Transmission interconnections between systems, coupled with internal system transmission facilities, allow for the economic exchange of electric power among all systems and industry participants. Such economy transfers help to reduce the cost of electric supply to customers."

[13] NERC Planning Standards I. System Adequacy and Security—Discussion.

While some also stress the requirement to plan a transmission system that will facilitate the electric market, others stress the need to focus on providing adequate reliability at minimum cost. While the goal of designing a system to facilitate a wholesale power market is commendable, the means for doing so have become extremely complex due, in large part, to structural changes being implemented to establish the markets.

Transmission planning used to involve dealing with a more or less consistent pattern of power flow from known sources to known load pockets. Even then, in many areas the time it took to plan and get approval for new transmission lines could take many years, even when there were relatively few regulatory jurisdictions involved. Going forward the level of uncertainty has increased dramatically. Uncertainties are of both a technical nature and of a structural/financial nature. Among them are:

- How much transmission capability is needed, in both directions, across each and every existing and future interface and flowgate?
- How is the magnitude of this capability determined since:
  - In spite of the many numerous proposals to build merchant power plants, the likelihood is that many will never be built.
  - The locations and sizes and dispatch patterns of those that will be built are unknown.
  - The future of existing power plants may be unknown.
  - The impact of dispersed generation is unknown.
  - Of those power plants built, how will they be dispatched, given the commercial sensitivity of the contractual arrangement that they will enter into for sale of the power or the bidding strategies they will use?
- How will variations in the assumed or best-guess generation expansion plan impact a transmission development plan?
- How will trade-offs between different generation options, FACTS device applications, and load-curtailment measures be identified and evaluated given the different financial perspectives of the various involved parties?
- Is there any potential for realizing economies-of-scale, that is, building a higher voltage line initially instead of a series of lower voltage/capacity lines?
- Can lines ever be built solely for reliability purposes?
- Who will finance and build the transmission?
- What will be the rules for paying for use of the transmission?
- What types return on transmission investments will be needed?
- Given the uncertainties involved, how can a plan gain the requisite regulatory approvals, especially if multiple jurisdictions are involved?

9

RELIABILITY

Interruptions in the supply of electricity to customers can occur at any hour of the day or night and can last from fractions of a second to many hours or even days. Interruptions can be caused by disturbances to or malfunctions of any of the three components of the power system—(1) generation, (2) transmission, or (3) distribution. They can also be caused by the unavailability of adequate resources to supply the customer load. These two attributes of reliability are characterized by NERC as security and adequacy.

Data show that over 90% of customer outages are caused by problems originating on the local distribution system. Although generation and transmission related outages are less common than those related to the distribution system, they often have much more serious consequences because of the number of customers affected and the duration of the outage.

Disturbances can be initiated by:

- External events such as:
  - Environmental factors such as wind, rain, lightening, ice, fire, floods, earthquakes;
  - Accidents such as cars hitting poles; and sadly
  - Sabotage.
- Internal events such as:
  - Insufficient capacity;

Understanding Electric Power Systems: An Overview of the Technology and the Marketplace, by Jack Casazza and Frank Delea
ISBN 0-471-44652-1

- Failure of equipment due to electrical or mechanical stresses;
- Operating errors or decisions.

Lack of resources can be due to:

- Insufficient generation caused by:
  - Low load forecasts;
  - Shortages of fuel due to supply disruptions or delivery/transportation problems;
  - Opposition to the construction of required new generating capacity;
  - Failure of equipment due to electrical or mechanical stresses;
  - Poor planning;
  - Excessive maintenance outages;
  - Regulatory actions restricting the operation of power plants;
  - Transmission constraints;
  - Generation being retired because it is non-competitive in the new competitive market.
- Insufficient transmission or distribution caused by:
  - Low load forecasts;
  - Opposition to the construction of required new transmission or distribution lines;
  - Failure of equipment due to electrical or mechanical stresses;
  - Poor planning;
  - Intentional outages required because of other infrastructure work, that is, the widening of roads.

The duration of the interruption will be affected by the severity of the disturbance, the power system facilities affected, the redundancy or reserve built into the system, and the preparedness of the involved operating entities to respond. Some interruptions are of very short duration because the disturbance is transient and the system self corrects. Some interruptions, such as those caused by tornadoes or ice storms, damage significant portions of the system requiring many days to restore service. When there are insufficient generation resources, the outages may be of a controlled and rotating nature. Their duration might be only during peak load hours.

The extent of the interruption will be determined by the initiating disturbances and the facilities affected. For example, cascading outages caused by a fault occurring when a system is operating above a safe level can involve many states as can a widespread ice storm. Conversely, a distribution pole damaged by a car may affect only a few homes.

An increasingly important aspect of power system reliability is the quality of service or power quality. With the increasing importance of computers and new electronic communication procedures, imperfections in electric service become increasingly important to the customer. Such imperfections include:

- Momentary interruptions;
- Voltages outside of acceptable limits;
- Voltage dips of very short duration.

Protection against power quality imperfections can often be handled by the consumer. Pressure is mounting, however, for the supplier to improve quality. This raises the question of the responsibility for such improvements in an unbundled power industry with separate companies providing distribution, transmission, and power supply services.

COSTS OF POWER OUTAGES

The costs of electric power outages to American electric customers are generally called "socio-economic" costs. Attempts have been made to quantify these costs but the estimates vary widely. One source reports that the costs are $26 billion each year and that they have been increasing as the electric power industry is restructured. A 2001 report[1] from the Electric Power Research Institute (EPRI) states that power outages and problems with power quality cost the U.S. economy over $119 billion per year.

Numerous impacts of power outages have been identified. Included among these impacts are:

- Loss of life due to accidents (e.g., no street lights);
- Loss of life of ill and elderly (death rates go up);
- Loss of productivity by industry;
- Loss of sales by business;
- Loss of wages of labor;
- Damage to equipment in industry;
- Fires and explosions;
- Riots and thefts;
- Increased insurance rates.

[1] "The Cost of Power Disturbances to Industrial and Digital Economy Companies," June 2001. EPRI.

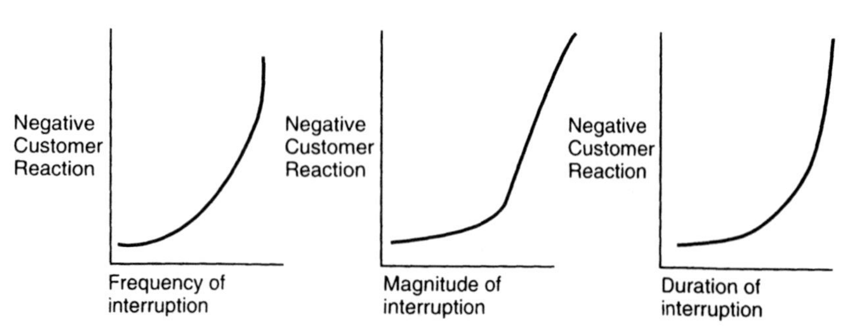

Figure 9.1. Consumer reaction to interruptions.

WAYS TO MEASURE RELIABILITY

Reliability of a system is difficult to measure. Perhaps the best way is through evaluation of the consequences of possible consumer interruptions. Investigations have shown that the best measure of reliability is that of consumer reaction.

Five conditions that have been identified impact the value an average consumer puts on an unsupplied mWh of lost energy:[2]

1. The activities affected by the curtailment and therefore the time of day and mix of customers;
2. The number of interruptions;
3. Availability of advance warning;
4. Weather conditions and therefore the time of year;
5. The duration of the interruption.

Figure 9.1 shows that this reaction increases dramatically as the frequency of outages increase, as the duration of the outage increases, and with the magnitude or extent of the outage. The following function presents a means of evaluating this reaction:[3]

$$R = \text{function of } \{K, F, T, P, t\},$$

where K is an empirical coefficient proportional to the consumer's dependence on electricity; F is the frequency of interruptions; T is equal to duration of the interruptions; P is the amount of load interrupted, and t is the time when the interruption occurs.

[2] Crampton and Lein, *Value of Lost Load*, University of Maryland, February 2000.

[3] J.A. Casazza, *Generation and Transmission Reliability*, CIGRE Paper #32-11, Paris, France, 1970.

Experience has shown that K increases with increasing consumption of electricity per customer, and t is greatest at the time of day, week, or year when people suffer the greatest hardships if service is interrupted. This criterion for reliability evaluation does not consider other curtailments of service, such as voltage or frequency reductions. These "partial" curtailments are not as important to most consumers as a complete interruption but they should also be considered.

A number of indices have been developed, primarily for the distribution system, to provide another measure of reliability:

SAIFI System Average Interruption Frequency Index; measures the average frequency of sustained interruptions per customer

SAIDI System Average Interruption Duration Index; measures the average time that all customers are interrupted

CAIDI Customer Average Interruption Duration Index; represents the average time required to restore service to the average customer per sustained outage

MAIFI Momentary Average Interruption Frequency Index; tracks the average frequency of momentary interruptions, typically defined as less than five minutes

PLANNING AND OPERATING A RELIABLE AND ADEQUATE POWER SYSTEM

As discussed in Chapter 8 the electric utility industry, over time, developed planning, operating, and design standards to address customer expectations of reliable service. These standards were at first local in perspective but, as interties were built and the interdependent nature of the system became apparent, many of the standards were expanded to a regional and then a national perspective.

Concurrently, over the last century and continuing to the present, customer expectations of reliable service have also increased. Outages which once were common place are now considered unacceptable. Momentary interruptions, which at one time were noticed by only a few customers, now impact many customers because of the widespread use of computers and other electronic devices.

Underlying the industry's approach to reliability was the realization that its efforts should be multi-dimensional:

- Plan the system to have enough generation, transmission, and distribution capacity.
- Design the system to reduce the probability of equipment failure.
- Operate the system to remain within safe operating margins.

- Be prepared to restore the system quickly, in the event of a supply disruption.

In all cases, the industries' efforts involve a trade-off between reliability and costs. It would be impossible to build enough facilities or operate with enough of a reserve margin to have a perfectly reliable system. For example, some types of common-mode failures due to causes such as tornadoes, ice storms, or hurricanes involve so many facilities that it would be financially impossible to design a system to tolerate them. This is why the requirement for restoration plans is so important.

The bases for the standards that have been developed are varied. All reflect, in one way or another, a view as to an acceptable level of reliability. Generation planning standards have, in the past, been tied to a statistical measure. Standards for operating the generation and transmission systems are based primarily on the collective judgments of utility personnel. Over the years, these standards have been accepted and legitimized by local, state, and national regulators in rate cases and in after-the-fact reviews of outages. In many of these reviews, customer complaints over service reliability and over costs have caused modifications to aspects of individual standards. For example, problems in some areas with long restoration times after major storms have led to requirements for detailed and publicized restoration plans reflecting customer inputs.

Many of the industry standards, especially those that relate to the bulk power system, are being assembled and made applicable at the national level by NERC and its committees and working groups.[4] This effort reflects the expanded number of participants active in the industry.

Transmission Security and Security Coordinators

NERC policies for dealing with transmission security require that control areas, individually and jointly, shall develop, maintain, and implement formal policies and procedures to provide for transmission security. These policies and procedures shall address the execution and coordination of activities that impact inter- and -intra-Regional security, including:

- Equipment ratings;
- Monitoring and controlling voltage levels and real and reactive power flows;
- Switching transmission elements;
- Planned outages of transmission elements;
- Development of Operating Security Limits;
- Responding to operating security limit violations:

[4] See www.nerc.com

- Responsibility for transmission security. When operating security limit violations occur, or are expected to occur, the control areas affected by and the control areas contributing to these violations shall implement established joint actions to restore transmission security.
- Action to keep transmission within limits. Control areas shall take all appropriate action up to and including shedding of firm load in order to comply with Standard 2.A.2.

They also require that every region, sub-region, or inter-regional coordinating group shall establish one or more security coordinators to continuously assess transmission security and coordinate emergency operations among the control areas within the sub-region, region, and across the regional boundaries. Transmission operating entities are required to cooperate with their host control areas to ensure their operations support the reliability of the interconnection. This includes coordination in the planning of transmission outages with any system that operations planning studies show may be affected.

Information on reliability issues is also obtained from post incident reviews covering major regional disturbances both in the United States and overseas. Since the laws of electricity apply to all transmission grids, lessons learned from overseas blackouts can fruitfully be applied in the United States.

Some specific examples of major disturbances where lessons were learned are the reports on:

- 1965—Northeast United States blackout;
- 1967—Mid-Atlantic Blackout;
- 1977—New York City Blackout;
- 1978—France Blackout;
- 1997—California Blackouts;
- 1997—New Zealand Blackouts.

In some instances the lessons learned are technical in nature, that is, the criticality of voltage support as demonstrated in the French Blackout of 1978. In other instances the lessons are organizational, that is, the need for expanded planning and operational coordination after the great Northeast Blackout of 1965 and the need for better intra and inter-area communication after the New York City Blackout of 1977.

As the structure of the industry evolves, the trade-offs between reliability and cost will become more difficult, in that different companies will be involved in the supply-delivery chain, and each will have a different perspective on the cost—benefit relationship of reliability. Reliability will depend on whether the "three musketeers" or the "lone ranger" approach is used. With the "three musketeers" approach, problems of one system or company are shared by all in an effort to minimize total societal costs. In the "lone ranger" approach, each system or business customers suffer alone the consequences of its problems. Some believe this will provide motivation for all to meet their obligations.

Paying for Extra Reliability

One finds in the literature discussions of the customer's willingness to pay for a greater level of reliability. There are two ways to give greater levels of service to individual customers:

1. Provide more redundancy of supply to one customer than to another.
2. In the event of a disturbance or insufficient capacity, disconnect or interrupt the customer who doesn't pay a premium rate for electricity.

Given the reality of how a power system is physically structured, the redundancy option has limited application in protecting specific customers against transmission facility outages especially when the exposure is to a security violation, that is, loss of a facility. In select circumstances, a larger customer may be able to have a higher level of local distribution service by providing him with another distribution feeder or transformer, but extending the option to the typical customer would become cost prohibitive if individual distribution facilities were to be targeted to individual customers. The same logic applies to the transmission grid. Additionally, trying to distinguish between customers at the transmission level during dynamic conditions where instability occurs would be impossible under many conditions.

If the reliability problem is one of adequacy, that is, insufficient resources—when operating personnel have time to take corrective action, customers willing to pay a higher rate could be given preference when adjustments have to be made to restore the load—generation balance. At the generation level, individual customers could arrange that their supplier maintain additional generating reserves for them at added cost. The process of implementing such a plan could rely on either financial mechanisms or physical mechanisms to disconnect customers not opting for higher levels of reliability.

Compliance

In the past, compliance with national reliability standards has relied on the voluntary cooperation of the companies involved. Procedures are being put in place in some regions for implementation of financial penalties for noncompliance. Recognizing the changing nature of the industry, NERC, the National Association of Utility Commissioner's (NARUC), and numerous other organizations are pressing Congress for legislation mandating compliance with those standards developed for the bulk power system under the auspices of NERC.

GENERATION

As discussed earlier, generation adequacy standards are of two types: those covering the amount of installed generation capacity and those covering the amount of operating generating capacity. Generating capacity has traditionally been installed to meet a statistically determined reserve requirement, that is, an amount of installed capacity over and above the expected peak load obligation of the supplier. The amount of required reserve was related to a probably of loss of load. The precise determination was tailored to each system and reflected its planning and operating philosophies.

In the evolving industry, the question is unanswered of whether the level of installed generation capacity should be a design requirement or should be market determined. NERC has removed from its planning criteria a requirement for a targeted installed reserve. NARUC as part of its National Electric policy states:

> "Congress should mandate compliance with industry-developed reliability standards on the bulk power system that *includes adequate reserve margins* (emphasis added) and preserves the authority of the States to set more rigorous standards when deemed to be in the public interest."

In addition to an installed generation reserve, an operating reserve is also defined. On a day-to-day and hour-to-hour basis, there must at all times be sufficient generation synchronized to the grid to meet the load requirements at that time and to be able to respond to short-term variations in customer load, as well as to cover for the loss of another generator. The industry has rules governing the amount of operation reserve required at any one time. The reserve is usually related to the amount of generation that could be lost as a result of any one of the group of disturbances. This operating reserve must be available within 15 minutes and is usually allocated among groups of companies who have agreed to jointly maintain these reserves. The allocation of responsibility to each company is then assigned to individual generating units. Allowance is made to reduce load also as part of operating reserves.

These reserves and this redundancy must cover both the real and reactive needs of the system. If insufficient reactive is available, voltages on the system will decline and the system could suffer a cascading power interruption which results in a complete blackout.[5]

Another important consideration in the installed generation picture is the diversity of fuel supply. Consistent with costs, a fuel diverse generation mix provides an additional level of reliability. As examples, hydro systems are exposed to the impact of droughts; while coal and oil fired systems can be

[5] A good example of this phenomenon occurred in December 1978, in France, and resulted in the power supply for the entire nation being shut down.

impacted by a number of disruptions including worker strikes, disruption in boat deliveries of fuel, freezing of coal piles in the winter, etc.

TRANSMISSION

Transmission System Problems

Transmission systems are aging and rapidly growing less adequate. The average age for transmission lines, transmission cables, circuit breakers, switchgear, substations, transformers, and other equipment is approaching 30 years, with some key facilities more than 75 years old. Needed maintenance has not been performed in recent years in order to cut costs and improve profits, and maintenance needs are growing rapidly. Such maintenance requires equipment to be taken out of service. Transmission outages on our existing systems can be expected to continue to increase. It is growing increasingly difficult to schedule such outages without taking large reliability risks or incurring large cost penalties due to the inability to deliver low-cost power.

The ability to develop a reliable transmission system has decreased sharply in recent years as generators are added and retired from service. Transmission system loading conditions and capacity are affected every time a generator unit is added or retired. Plans exist for the addition of many new units. Purchasers of existing plants expect to retire those that are not sufficiently profitable or where they wish to use the plant site for other purposes. Many of these plans are being made without recognizing the impact on the transmission system. The time needed to reinforce the transmission system is often much longer than the time required to make generation additions. Public opposition to new transmission lines cannot be expected to decrease. Generator retirements can be made at any time by the plant owners with very little notification. There are no provisions for the transmission reinforcements that may be needed because of retirements or who will pay for them.

Capital expenditure reductions, reduced maintenance expenditures, increased transmission outages, lack of time to maintain and reinforce the system are all occurring. Needed new transmission lines have not been added in recent years. (An average of about 13,000 miles of transmission were added in the United States in the 1990–1992 period, while the average in the 1996–1999 period was about 6,000.) This trend is creating the potential for a national disaster.

A basic question is, "Can market forces result in an economical and reliable transmission system?" In the United States we have three huge synchronous systems (Figure 1.3). Each of these synchronous systems is a "single machine". This means that outages, generation and transmission changes, problems in any one area in the synchronous network can affect the entire network. Changes in one location affect not only line loadings and voltages, but also stability limits, short-circuit duties, and required relaying in other systems. Problems in California affect the northwest and Arizona. If a generator is lost in New York City, its effect is felt in Georgia, Florida, Minneapolis, St. Louis, and New Orleans. Transmission lines cannot be added helter-skelter based solely on the profits for the owner. Locations and designs for new substations selected by the distribution systems must recognize the future of the transmission system which will supply them. The transmission system must be designed as an integrated whole. One cannot design a reliable low-cost automobile by having separate uncoordinated designs for the brakes, the transmission, the engine, and other essential systems. The same is true for the electrical transmission system. It must be designed as an integrated whole.

Transmission systems must also be optimized in "three dimensions" in order to achieve necessary reliability and minimum costs for electric power. They should be optimized "geographically", that is, the transmission system must meet the needs of all who are served by the synchronous network, not just the needs or the profits of any one system, any one area, or any one region. They must be optimized "functionally", that is, the transmission system must meet both generation requirements and the requirements of the distribution systems which they supply. These requirements must be balanced on an overall basis. And lastly, transmission systems must be developed to meet needs over a significant period of time since they cannot be changed once constructed. We must develop transmission systems that not only meet needs this year, but next year, and five years down the road. We must develop transmission systems that also recognize our long-term needs. They must be optimized "chronologically".

Can market forces be used to develop such optimized long-range transmission systems? Under current procedures the locations of planned generator additions are not known beyond one or two years in the future. Many believe that the only way to develop a transmission system that meets the needs of this nation is through a National Power Survey which reviews our long-term future needs and the transmission systems have been proposed on a national basis similar to past National Power Surveys.

Planning and Operating Standards

The planning and operating standards for the bulk power system are the most national in nature. Originally they were developed by individual utilities, afterwards they were established on a regional council and then on a national basis under NERC. We've discussed elements of these standards in Chapter 8.

Attempts have been made to determine and set the level of transmission system reliability based on the reliability of each of the components of the system. Although in theory appealing, this effort flounders on the magnitude and variations in equipment that constitute a power system. The system is designed to reflect good engineering judgment. For example, an engineer can select a number of designs for a new bulk power substation depending on its criticality. The planner could select a substation with a breaker and a half arrangement which provides more redundancy and, hence, a higher level of reliability than a ring bus design provides.

Since the effects of electrical disturbances can spread over a wide, multistate region, the need for regional coordination in planning and operation is obvious. As the new market rules for the electric system are developed, the concern is that the rules in any one area do not lower the local reliability standards and thereby impact or impair the reliability of the grid.

Voltage and Reactive Control

Present NERC reliability standards require:

1. *Monitoring and controlling voltage and mVAr flows.* Each control area individually or jointly, shall ensure that formal policies and procedures are developed, maintained, and implemented for monitoring and controlling voltage levels and mVAr flows within its boundaries and with neighboring control areas.
2. *Providing reactive resources.* Each control area shall supply reactive resources within its boundaries to protect the voltage levels under contingency conditions. This includes the control area's share of the reactive requirements of interconnecting transmission circuits.
   a. 2.1 Providing for reactive requirements. Each purchasing selling entity shall arrange for (self-provide or purchase) reactive resources for its reactive requirements.
3. *Operating reactive resources.* Each control area shall operate their capacitive and inductive reactive resources to maintain system and interconnection voltages within established limits.
   a. 3.1 Actions. Reactive generation scheduling, transmission line and reactive resource switching, etc., and load shedding, if necessary, shall be implemented to maintain these voltage levels.
   b. 3.2 Reactive Resources. Each control area shall maintain reactive resources to support its voltage under first contingency conditions.
   c. 3.2.1 Location. Reactive resources shall be dispersed and located electrically so that they can be applied effectively and quickly when contingencies occur.
   d. 3.2.2 Reactive Restoration. When a generator's voltage regulator is out of service, field excitation shall be maintained at a level to maintain interconnection and generator stability.
4. *Providing operator information.* The system operator shall be provided information on all available generation and transmission reactive power resources, including the status of voltage regulators and power system stabilizers.

5. *Preventing Voltage Collapse.* The system operator shall take corrective action, including load reduction, necessary to prevent voltage collapse when reactive resources are insufficient.
6. *Providing voltage and reactive devices.* Devices used to regulate transmission voltage and reactive flow shall be available under the direction of the system operator.

DISTRIBUTION

Planning and operation of the distribution system is still done according to the standards and practices of individual utilities and reflect local reliability requirements and cost considerations. The robustness of the supply to a congested urban area will be considerably greater than that to a rural farm district. However, oversight of the local utility's performance is usually exercised by state regulatory authorities. It's not uncommon for post-incident reviews to be held by regulators after significant local outages. In many areas, the use of incentive rates-of-return reflecting distribution system performance is becoming popular. Utility design practices reflect standards developed by national organizations such as the Institute of Electrical and Electronics Engineers (IEEE).

SUMMARY

Future reliability conditions on electric power systems are not subject to exact analyses. Load conditions on electric power systems vary continuously as customer utilization apparatus is switched on and off. As the loads vary and as supply equipment on the system is removed because of the need for maintenance or because of failure, identical conditions will not exist for two of the 8,760 hours in a year. While statistics can be accumulated, many other factors must be considered. By far the best individuals or organizations to make estimates of future reliability conditions are those most familiar with these factors. They must be close to what is going on, able to estimate future conditions and judge the sureness of the estimates involved, and to assess the relative risks of alternative courses of action. Reliable determinations are relative and best made by those with great experience.

10

RESTRUCTURING, COMPETITION AND DEREGULATION

CAUSES OF RESTRUCTURING

The electric power industry functioned under federal, state, and local regulation since the 1910s. From time to time events occurred which demonstrated that improvements were needed. Utility management at times failed to stress the need to minimize costs, with some having the attitude, "Don't worry about it, it will go in rate base and increase our earnings." Regulators often lacked the competence to understand the technology, and the system economics, and were more concerned with political results than the public welfare. In some states being chair of the public utility commission was the path to the Governorship.

In response to increasing costs of electricity, especially complaints from industrial customers, legislators and regulators looked to competition, the holy grail of restructuring! It could be introduced by giving customers a choice of supplier. Large industrials such as aluminum companies, chlorine companies, and so on, could chose locations for their future plants based on rates offered by the companies. Organizations such as the railroads in the northeast, which had their own transmission systems that ran along their tracks, could negotiate to buy power from various utilities along the way. The goal of those favoring restructuring was to increase this competition by providing a choice of supplier to all users of electricity.

From the utility viewpoint, restructuring offered the opportunity for getting out from under often incompetent and unfair regulation. From a large indus-

Understanding Electric Power Systems: An Overview of the Technology and the Marketplace, by Jack Casazza and Frank Delea
ISBN 0-471-44652-1

trial user viewpoint, there was a chance to obtain a fairer rate structure and perhaps lower rates through competition. Lastly, many professionals, lawyers, accountants, engineers, company executives, saw an opportunity to gain considerable additional income as a result of restructuring.

The restructuring effort needed to consider the unusual characteristics of electric power systems as discussed in previous chapters. Electricity cannot be stored, electricity flows as dictated by Kirchoff's laws and not contracts, and decisions by one supplier to provide or not provide certain services could affect the costs of others connected to the electric power system and the total cost of electricity.

TYPES OF RESTRUCTURING

Worldwide, three basic types of institutional changes occurred:

1. Privatization;
2. Introduction of new competitive procedures;
3. New forms of regulation.

In the United States, the majority of the electric supply was from utilities that were already privately owned and the major changes were in competitive procedures and forms of regulation. Suggestions to privatize government-owned systems as the PMA's, TVA, Bonneville, and municipals met with strong political opposition, and were soon dropped. Those supplied by these governmental systems would lose their tax and financing advantages, and their costs would increase significantly.

Of particular interest are the new competitive procedures that have been put into place:

1. The unbundling of generation, transmission, and distribution into separate businesses.
2. Open transmission access.
3. Control of the operation of the United States' transmission system assigned to a relatively few organizations.
4. Creation of a new class of non-regulated generators.
5. The formation of organizations for dispatch of all generation.
6. Prices for bulk power supply ceased to be cost based.
7. Competition in power production was achieved with dispatch based on quoted prices.
8. The construction of many gas-burning power plants by independent owners.
9. Payments for some power production based on 'market clearing' price, which is the highest of any accepted bid prices.

10. Retail wheeling increasing in stages to smaller and smaller size customers.
11. No central planning.
12. Hedging contracts to guarantee prices to purchasers in a volatile market.

Effects of Restructuring

Some of the effects of these new competitive mechanisms were:

1. Bulk power prices for electricity became exceedingly volatile, changing very drastically from one day to the next, and reaching levels undreamt of when the new markets were established.
2. The volatility of this pricing fostered the development of organizations that offer 'hedging' contracts. These contracts guarantee the price of electricity to the purchaser thus reducing a cost uncertainty in their operation. This is a form of insurance but, like all other forms of insurance, this has added to the cost of electricity a component that did not exist under the previous institutional structure.
3. The driving force for the operation of the various unbundled companies has been based on maximizing their own profit, not on minimizing electricity costs.

Many believe that the new competitive mechanisms introduced have caused very significant cost increases that have harmed the users of electricity. While improvements are being made to correct past errors, there is increasing emphasis on an increase in regulatory control, centered more on a national or regional basis than on state-by-state basis as in the past.

SIX NETWORKS

An understanding of the electric power industry and of the implications of electric power policy can be achieved by examining six networks describing the industry. The networks are:

1. *The physical network*, which supplies electric power consisting of generators, transformers, circuit breakers, transmission lines, conductors, substations, and so forth. This network can be drawn as a diagram which is used in its analysis. This network has been described in the previous chapters of this book.
2. *The fuel/energy network*, which includes all sources of energy and fuel which are converted by various means, delivered by various means, and utilized by various means. It has been described in Chapter 5. This

network can be diagrammed and used for analysis.[1] The energy network supplies the physical electric network in various ways. Examination of the various paths of fuel delivery and their efficiencies is helpful in understanding how all energy is consumed. For example, the losses in electric transmission and distribution can be compared with the losses and energy required to ship gas through pipelines, and so on. The losses in the energy network are a significant portion of the total energy resources consumed and require far more attention.

3. *The regulatory network*, by which the government controls the power industry. In the case of the United States, this includes regulation by federal government, the state governments, and by local municipalities. The regulatory network controls pricing, reviews and approves various aspects of system expansion, and establishes controls on the contractual arrangements between the various business participants.
4. *The business network*, which includes the ownership of the facilities involved and the contracts between the various parties. An important part of this network is the tariff or pricing arrangements established by the parties, usually in conformance of the restraints of the regulatory networks.
5. *The money network*, which includes the money sources and the flow of money among the various participants. Money sources are analogous to generators in the electric power system. The money sources are the stockholders, consumers, taxpayers, banks, and so forth. By diagramming the sources of money, its flow through the networks, and its use for various functions, one can get a far better understanding of how our financial systems work. These money networks have been drawn in the past, starting in the early 1960s initially by engineers. There are transformation points, or coupling points, between the money network, the energy network, and the power network. Money is provided to and received from the other networks.
6. *The information, communications, and control network*, through which information is transferred. This information is used as part of the operation of the power market for e-commerce, for scheduling and dispatching the power system and the control of the power system and other energy systems.

It is by understanding the operation of these networks and how the operation of one affects the others that one can begin to comprehend the affects of various policies.

[1] "Reference Energy Systems and Resource Data for Use in Assessment of Energy Technologies" Associated Universities of Office of Science and Technology, April 1972.

CHANGING CUSTOMER REQUIREMENTS

As pointed out by Michael Zimmer,

> By 2006, the U.S. Department of Commerce estimated that almost half of the U.S. workforce will be employed by industries that are either major producers or intensive users of information technology and e-commerce products and services. Information technology and e-commerce industries have contributed over one-third of the Nation's real economic growth from 1998–1999. The one underlying constant is that these industries are powered with electricity for many processes, marketing, storage, and operating facilities. Without electricity, provided reliable and consistent quality levels, these technology industries simply cannot operate. The continuing transformation of the electricity industry may offer advantages to the Internet and information industries, but it may also offer significant challenges to meet future growth, pricing, efficiency, and reliability standards for service for these information-based companies. Will quality and service decline? Will reliability management and responsibility shift? Will decentralized systems thrive as utility service is altered?

Increasingly, customers are asking for the ability to obtain from their electricity supplier a higher than normal reliability or quality of service. Claims by some that the utility system should be able to provide such service are ignorant of the basic functioning of electric power systems as described previously in this book. All customers are served by the same transmission system.

The ability of a restructured competitive electric power industry to change overall system designs to meet these future requirements for improved quality raises key questions. The quality of service depends on coordination of the various participants in the system, many of whom are competitors. The requirement for improving the power supply quality may prove an important incentive for the development of many new decentralized power sources, resulting in additional industry restructuring changes.

11

LEGISLATION AND REGULATION—THE REGULATORY NETWORK

The physical and fuel energy networks have been described in prior chapters. A third, and equally important network, is the regulatory network. The functioning of this network is completely controlled by human beings and human decisions. It plays a vital role in the functioning of all the other networks, sometimes providing specific rules for their functioning while at other times providing restraints within which their operation must be conducted. From time-to-time failures to recognize the need for the regulatory networks to be consistent with the physics and operational requirements of the other networks has resulted in problems which have increased costs and decreased the reliability of electric power systems. The results in California and many other regions in the United States provide outstanding examples.

PRICING AND REGULATION

The reconciliation of the interests of the private ownership of the facilities and systems that provide electric power and the public interest character of their business is accomplished by the concept of regulation. Regulation refers to the laws, and the actions of tribunals established under those laws, governing the business of utilities.

The concept of regulation of private business affected with a public interest is not a new one. As long ago as during the reign of the Roman Emperor Justinian, laws were passed regulating the business of privately owned docking

Understanding Electric Power Systems: An Overview of the Technology and the Marketplace, by Jack Casazza and Frank Delea
ISBN 0-471-44652-1

facilities assuring access to all interested individuals. There were a limited number of choice sites for docks and, once those were occupied, other commercial interests were effectively denied access which was recognized as detrimental to the necessary commercial activities in the Empire. Since that time, granaries, grist mills, breweries, thoroughfares, and a number of other activities have been considered so important to the public interest that special laws were passed regulating such businesses.

Electricity providers are regulated at the federal, state, and local levels. Every state has a regulatory agency directly concerned with the activities of utilities. The federal government has several agencies supervising various aspects of utilities. Local municipalities regulate land use and, in some cases, certain franchise aspects.

In the past, the principal regulation of electric utilities was by state regulatory authorities. Federal regulation was generally limited to activities involving interstate commerce or situations in which the national interest was affected.

FEDERAL LEGISLATION

The foundation of federal regulation of electric utilities is the Public Utilities Holding Company Act of 1935 (PUHCA) and the Federal Power Act (FPA).

Public Utility Holding Company Act of 1935

At the turn of the century, America was dotted many small electric utilities, most of which were isolated plants serving small areas within existing cities. Entrepreneurs began buying up these small utilities to form larger and larger systems.

To construct plants and lines, utility owners had to raise substantial capital from the securities markets. Entrepreneurs used relatively little of their own capital to control utilities, by purchasing with non-voting debt and only a small portion of equity. Holding companies provided a means of extending this method of leveraging small amounts of capital into ever-larger control across many firms. Through this approach one holding company, the Standard Gas and Electric Company, was able to control assets worth $1.2 billion with an equity investment of only $23,000. Companies of unprecedented scope were created. These holding companies produced almost half of the electricity produced in the United States.

A history of financial abuse was associated with these large holding companies. The abuses included pyramid financial structures with very high levels of debt, inadequate disclosure of the financial position and earning power of holding companies, unsound accounting practices, and abusive affiliate transactions. Economic benefits did result, however, through the advantages of the economy of scale and coordination of systems geographically and functionality. They provided better engineering and facilitated the raising of capital. By the 1930s reactions to the excessive power and abuses of these large companies lead to the passage of the 1935 Holding Company Act (PUHCA).

PUHCA states that:

> "Interstate holding companies, engaged through subsidiaries, in the electric utility business or in the retail distribution of natural or manufactured gas are subject to regulation under this Act. These companies, unless specifically exempted, are required to submit reports providing detailed information concerning the subsidiaries. Holding companies are subject to the SEC regulations on matters such as structure of their utility systems, transactions among companies that are part of the holding company utility system, acquisitions, business combinations, the issue and sale of securities, and financial transactions."[1]

> "The 1935 Act addressed these problems by giving the Commission authority over various practices of holding companies, including their issuance of securities and their ability to engage in affiliate transactions. The Act also placed restrictions on the geographic scope of holding company systems and limited registered holding companies to activities related to their gas or electric businesses."

In the years following the passage of the 1935 Act, the SEC worked to reorganize and simplify existing public utility holding companies in order to eliminate abuses.[2] As a result of PHUCA, there was a large decrease in the number of utility holding companies. By 1958 there were only 18, as compared to 216 in 1938.[3]

PUHCA effectively "closed the door" on new entrants to the private utility electric generation business since any entity owning 10% or more of a utility had to divest all of its non-utility assets. The private utility model became a vertically integrated utility which supplied generation, transmission, and distribution service in a specified area, that is, a franchise area.

Federal Power Act

The Federal Power Act of 1935 gave the Federal Power Commission ("FPC")[4] regulatory power over the terms, conditions, and rates of interstate and wholesale transactions and transmission of electric power to ensure electricity rates that are "reasonable, non-discriminatory, and just to the consumer."

[1] The full text of this Act is available on-line at *www.law.cornell.edu/uscode/15/ch2C.html*

[2] Testimony of Isaac C. Hunt, Jr., Commissioner, SEC before the Subcommittee on Energy and Air Quality, Committee on Energy and Commerce, U.S. House of Representatives, February 13, 2002.

[3] In the 1990s, due to, in large part to the EPACT of 1992, many new holding companies were formed. This issue will be explored in Chapter 12.

[4] The FPC had been established under the Federal Water Power Act of 1920 to encourage the development of hydroelectric power plants. The Commission originally consisted of the Secretaries of War, Interior, and Agriculture.

The FPA changed the structure of the FPC so that it consisted of five commissioners nominated by the President for five year terms, with the stipulation that no more than three commissioners could come from the same political party.[5] It directed the FPC to divide the country into regional districts for "voluntary interconnection and coordination". The Commission had the duty to promote and encourage such interconnection and coordination. The FPC also had the authority, when applied for by a utility, to order another utility to physically interconnect and to sell or exchange energy.

Other Federal Laws

In the 1930s, other acts were passed which supported and encouraged the development of publicly owned utilities or the sale of power generated at power plants owned by governmental organizations:

- 1933—The Tennessee Valley Authority Act (TVA), which dealt with the provision of electric service in what were then rural areas. The TVA was authorized to generate, transmit, and sell electricity. It could build transmission lines as needed.
- 1936—The Rural Electrification Act (REA), which allowed loans to organizations providing electricity to sparsely populated rural areas.
- 1937—The Bonneville Power Act, which created the Bonneville Power Administration responsible for the transmission and marketing of power from Federally constructed dams.

Environmental Laws

Figure 11.1 lists the major Federal laws which deal with environmental matters, many of which impact electric utilities.

The National Environmental Policy Act (NEPA) requires federal agencies to integrate environmental values into their decision-making processes by considering the environmental impacts of their proposed actions and reasonable alternatives to those actions. To meet this requirement, federal agencies prepare a detailed statement known as the Environmental Impact Statement ("EIS").

The Clean Air Act is a comprehensive Federal law that regulates air emissions from area, stationary, and mobile sources. This law authorizes the U.S. Environmental Protection Agency to establish National Ambient Air Quality Standards (NAAQS) to protect public health and the environment. The goal of the Act was to set and achieve NAAQS in every state by 1975. The setting of maximum pollutant standards was coupled with directing the states to develop state implementation plans (SIPs) applicable to appropriate indus-

[5] From FERC website *www.ferc.gov/about*

- Clean Air Act (CAA)
- Clean Water Act (CWA)
- Emergency Planning & Community Right-To-Know Act (EPCRA)
- Endangered Species Act
- Federal Insecticide, Fungicide and Rodenticide Act (FIFRA)
- Freedom of Information Act (FOIA)
- National Environmental Policy Act (NEPA)
- Occupational Safety and Health Act (OSHA)
- Oil Pollution Act of 1990 (OPA)
- Pollution Prevention Act (PPA)
- Resource Conservation and Recovery Act (RCRA)
- Safe Drinking Water Act (SDWA)
- Comprehensive Environmental Response, Compensation and Liability Act (CERCLA or Superfund)
- Superfund Amendments and Reauthorization Act
- Toxic Substances Control Act (TSCA)

Figure 11.1. Major environmental laws.
Source: *www.epa.gov*

trial sources in the state. The Act was amended in 1977 primarily to set new goals (dates) for achieving attainment of NAAQS since many areas of the country had failed to meet the deadlines.

> On January 1, 2000, the electric industry came under Phase II regulations of the Clean Air Act Amendments of 1990. This Act was primarily designed to reduce power plant emissions, specifically sulfur dioxide and nitrogen oxides. Phase I, which began on January 1, 1995, affected 435 generating units . . . Under Phase II, coverage increased to more than 2,000 units, . . . Since 1995, some generators have over-complied with Phase I in order to create excess allowances. This has allowed them to delay enacting additional strategies that would be necessary for compliance with Phase II. Strategies that are being used for compliance including fuel switching/blending, co-firing with natural gas, allowance acquisitions, scrubbers, repowering, and plant retirements.[6]

This environmental and other Federal legislation brought with it the concept of quantified emission rights. These rights can be bought and sold, thus facilitating meeting environmental requirements at minimum cost.

Department of Energy Organization Act

In 1977, the Department of Energy Organization Act created the Department of Energy. The law consolidated organizations from a dozen departments and agencies. Under this legislation, the FPC was replaced by the Federal Energy Regulation Commission ("FERC").

[6] This paragraph is from the EIA report Electric Power Annual 2000, Volume 1, August 2001.

PURPA

The Public Utilities Regulatory Act (PURPA) was one of a group of acts enacted in 1978 by Congress under the composite name of the National Energy Act (NEA). The NEA also included the National Energy Conservation Policy Act, the Power Plant and Industrial Fuel Use Act, the Energy Tax Act, and the Natural Gas Policy Act. Congress was responding to a number of issues that occurred during the 1970s which were impacting the cost or availability of power:

- The Arab oil embargo, which affected both the cost and security of the Nation's oil supply;
- A perceived developing shortage in the nation's proven reserves of natural gas[7];
- The rapid and steep increase in the cost of building nuclear power plants;
- The impact on power plants resulting from the Clean Air Act and other environmental legislation of the period;
- The slowdown in economic growth and its negative impact electric consumption, making many new power plants unneeded although the utilities were looking to recover their investments by raising their rates.

The many reasons cited in the literature for passing these laws can be combined into four main categories:

1. Lower the nation's oil and gas use.
2. Lower customer consumption by promoting efficiency.
3. Diversify the industry by promoting alternate energy technologies.
4. Lower costs to the consumer.

PURPA encouraged the development of alternative generation sources designated as "qualifying facilities" ("QFs"). There are two main types of QFs:

1. Cogenerators, which used a single fuel source to produce electric energy as well as another form of energy, such as heat or steam; and
2. Small power producers, which used renewable resources including solar, wind, biomass, geothermal, and hydro-electric power as their primary energy source. Each generator had to have a capacity less than 80mW.

QFs were exempted from regulation under PUHCA and the FERC so that the limitation on ownership contained in the PUHCA was bypassed.

Under Section 210 of PURPA, local utilities were required to purchase

[7] The shortages disappeared after price controls on natural gas were lifted in the 1980s and development began again.

power from QFs at a set price set by state public utility commissions, which was not to exceed the utility's avoided costs, and to sell back-up power to QFs. The amounts to be paid to the OFs were fixed for the length of the contracts. In many instances, the estimates of avoided costs used by some state commissions proved to be too high because of excessively high forecasts then prevailing of future oil prices. Subsequently, many utilities found themselves with obligations to buy power in the new marketplace at prices significantly above the pre-vailing levels. These contractual obligations were a major component of the stranded cost issue which affected many utilities' positions in the 1990s as the industry moved to a new posture on the ownership of wholesale generation.

Some states required utilities to enter into contracts with QFs, even when the utilities did not require the capacity. Recognizing these problems, in the late 1980s some states implemented competitive bidding procedures for required new capacity.

The development of OFs had mixed results in the short range. Additional generating resources were built using alternate or more efficient fuel but at the cost of steadily increasing prices for electricity. Regardless of the original rationale, PURPA's long-term effect was to introduce competition into the generation sector of the electricity market by creating a market for electricity produced by non-utility power producers.

PURPA also strengthened the FERC's power to order interconnections if it is in the public interest, or encourages conservation. The Commission was also given the authority to require a utility to provide transmission service (wheeling) to another utility. The Commission could order a utility to provide such service if it found such order:

"(1) is the public interest,
"(2) would—
  "(A) conserve a significant amount of energy,
  "(B) significantly promote the efficient use of facilities and resources, or
  "(C) improve the reliability of any electric utility system to which the order applies . . ."

Certain conditions had to be met before the Commission could order interconnection or wheeling. The conditions included:

- Does not cause uncompensated economic loss;
- Does not place an undue burden;
- Does not impair reliability;
- Does not impair ability to render adequate service.

The Commission could also exempt electric utilities from state laws, rules, or regulations which prevent voluntary coordination. PURPA also directed the FERC "study the opportunities for (a) conservation of energy, (b) optimization in the efficiency of use of facilities and resources, and (c) increased reliability, through pooling arrangements. The Commission could recommend to electric utilities that such utilities should voluntarily enter into negotiations where the opportunities for savings exist. The Commission was required to report annually to the President and Congress regarding any such recommendations and subsequent actions taken by electric utilities, by the Commission, and by the Secretary under this Act, the Federal Power Act, and any other provisions of law."

Energy Policy Act ("EPACT") of 1992

This wide reaching act contained further modifications to PUHCA and to the FPA intended to further increase competition in the generation sector.

PUHCA Modifications. PUHCA was modified to allow a new class of generation ownership, called Exempt Wholesale Generators (EWGs), to engage in the wholesale electric power market.

- EWGs are free from regulation under PURPA.
- Both registered and exempt holding companies can own EWGs.
- Utilities can buy electric energy from EWGs with which they are affiliated.

The law included provisions that require state regulators to review certain aspects of the financial implications of long term wholesale power contracts between utilities and EWGs. State PUCs have access to the "books, accounts, memoranda, contracts, and records of" any EWGs selling electricity to a utility subject to its jurisdiction, to the EWG, and to any associated company to the EWG including another utility or a holding company.

The act also allowed, subject to certain restrictions, exempt holding companies to purchase one or more foreign utilities.[8]

FPA Modifications. In order to provide a wide market for this new generation, the FPA was modified to require any utility providing transmission service to supply such service to anyone generating electric energy for sale for resale. The act required that the transmitting utility enlarge its transmission capacity where necessary to provide such services. The intent was to create an open-access transmission system. FERC was given authority to issue orders to implement this policy provided that the reliability of the electric systems affected by such an order would not be impaired.

[8] To enable investments by existing electric utilities in EWGs and to purchase or invest in foreign utilities, a large number of new holding companies were formed in the United States in the 1990s. Many utilities that purchased overseas utilities in the 1990s were forced to sell then in the early 2000s to reduce their high level of debt.

The law contained two perspectives on the use of the transmission system.[9]

1. The owners of the transmission were to be adequately recompensated for the use of the transmission; and
2. The rates charged should be economically efficient and . . . , just and reasonable and not unduly discriminatory.

FEDERAL REGULATORY AGENCIES

FERC[10]

The Federal Energy Regulatory Commission ("FERC") is an independent regulatory agency within the Department of Energy. FERC was created through the Department of Energy Organization Act in October 1977. At that time, its predecessor, the Federal Power Commission ("FPC") was abolished, and the new agency inherited most of the FPC's responsibilities. FERC:

- Regulates the transmission and wholesale sales of electricity in interstate commerce:
  - For private utilities, power marketers, power pools, power exchanges and independent system operators;
  - Reviews rates set by the Federal Power Marketing administrations, such as the Bonneville Power Administration;
  - Confers exempt wholesale generator status under the EPACT;
  - Certifies qualifying small power production and cogeneration facilities.
- Licenses and inspects private, municipal and state hydroelectric projects.
- Oversees environmental matters related to natural gas, oil, electricity, and hydroelectric projects.
- Regulates the transmission and sale of natural gas for resale in interstate commerce.
- Regulates the transmission of oil by pipeline in interstate commerce.
- Administers accounting and financial reporting regulations and conduct of jurisdictional companies.

[9] "Wholesale transmission services are to be provided at rates, charges, terms, and conditions which permit the recovery by such utility (the transmitting utility) of all of the costs incurred in connection with the transmission services, including, but not limited to, an appropriate share, if any, of legitimate, verifiable and economic costs, including taking into account any benefits to the transmission system of providing the transmission service, and the costs of any enlargement of transmission facilities. Such rates, charges, terms, and conditions, shall promote the economically efficient transmission and generation of electricity and shall be just and reasonable, and not unduly discriminatory or preferential." The Energy Policy Act of 1992, Title VII—Electricity, Subtitle B—Federal Power Act; Interstate Commerce of Electricity, Section 722, Transmission Services."

[10] The material in this section is from the FERC's Website: *www.ferc.gov*

- Oversees the issuance of certain stock and debt securities, assumption of obligations and liabilities, and mergers;
- Reviews the holding of officer and director positions between top officials in utilities and certain other firms they do business with, and;
- Approves site choices as well as abandonment of interstate pipeline facilities.

Its legal authority comes from the Federal Power Act of 1935 (FPA), the Natural Gas Act (NGA) of 1938, the Natural Gas Act (NGPA) of 1978, the Public Utility Regulatory Policies Act (PURPA) of 1978, and the Energy Policy Act (EPAct) of 1992.

SEC[11]

The SEC defines its role as: "The SEC also oversees other key participants in the securities world, including stock exchanges, broker-dealers, investment advisors, mutual funds, and *public utility holding companies* (emphasis added) . . . the SEC is concerned primarily with promoting disclosure of important information, enforcing the securities laws, and protecting investors who interact with these various organizations and individuals."

Its legal authority comes from the Public Utilities Holding Company Act (PUHCA) of 1935. Because of its role in addressing issues involving securities and financings, the SEC was charged with administering the Act.

Environmental Protection Agency (EPA)[12]

The EPA was established in July 1970 in response to the growing public demand for cleaner water, air, and land. "The mission of the U.S. Environmental Protection Agency is to protect human health and to safeguard the national environment—air, water, and land—upon which life depends."

In part, "EPA's purpose is to ensure that environmental protection is an integral consideration of U.S. policies concerning natural resources, human health, economic growth, energy (emphasis added), transportation, agriculture, industry, and international trade, and these factors are similarly considered in establishing environmental policy."

EPA focuses on three main areas regarding NEPA compliance:

1. Coordinating the review of all Environmental Impact Statements (EISs) prepared by other federal agencies;
2. Maintaining a national EIS filing system and publishing weekly notices of EISs available for review and summaries of EPA's comments; and

[11] The material in this section is from the SEC's Website: *www.sec.gov*

[12] The material in this section is from the EPA's Website: *www.epa.gov*

3. Assuring that EPA's own actions comply with NEPA and other environmental requirements.

Department of Energy (DOE)[13]

The Department of Energy's mission is to ensure national security. Some of the goals of three of DOE's four departments directly impacts the utility industry:

- The **energy program** to:
  - Increase domestic energy production;
  - Revolutionize our approach to energy conservation and efficiency;
  - Promote the development of renewable and alternative energy sources.
- The **environmental program** to:
  - Permanently and safely dispose of the nation's radioactive wastes.
- The **science program** to:
  - Sponsor cutting-edge science and technology research and development that revolutionizes how energy is found, produced, and delivered.

As part of its mission, the DOE has sponsored studies relating to the electric utility industry in a restructured environment.[14] Many experts consider these studies inadequate for setting future national policies.

Federal Legislation Under Consideration

As of early June 2003, Congress is debating the provisions of the Energy Policy Act of 2003. There are versions in both the House and the Senate. The House version of the bill has been approved by the House; the Senate's version is still under discussion. Once the Senate passes its version of the legislation, a committee of members of both houses will be formed to reconcile the different bill provisions. As with prior energy legislation, the proposals are far reaching covering numerous issues including, but not limited to, oil and gas, coal, nuclear energy, renewable energy, energy efficiency, research and development and electricity.

In May 2003, FERC's Chairman announced that FERC, at the request of the Senate Energy Committee Chairman, would wait until Congress took final action on the new legislation before it finalized its plan to implement the proposed SMD. This delay effectively delays implementation of the SMD until mid 2005, at the earliest.

[13] The material in this section is from the DOE's Website: *www.doe.gov*

[14] "Maintaining Reliability in a Competitive U.S. Electricity Industry, Final Report of the Task Force on Electric System Reliability September 29, 1998"; "The National Transmission Grid Study, U.S. Department of Energy May 2002."

STATE REGULATORY AUTHORITY

All states regulate the retail rates and tariffs of investor-owned utilities (IOUs). Some states regulate the rates and tariffs charged by municipal and cooperative utilities, generally on a more limited basis than for IOUs. Municipal utilities are an aspect of local government and the fact that the local electorate can exercise control over the policies of the municipal utility by its voting rights is considered generally to be an effective form of regulation not requiring further supervision by a state agency. The customers of cooperative utilities are its direct owners, and direct the activities of the utility by their vote.

Even so, a number of states in varying degree allow their regulating agency to supervise the rates and tariffs of municipal and cooperative utilities. In 20 states, the regulatory authority has at least limited jurisdiction over the rates of municipal utilities; in 30 states, the regulation of rates by the state agency extends to cooperatives.

As noted earlier, oversight of many federal environmental regulations was given to the individual states.

Many states consider some aspects of the Federal regulation and legislation currently (as of 2003) proposed as a major threat to their ability to control electricity costs and service reliability in their jurisdictions.

RECENT FEDERAL REGULATION IMPACTING THE ELECTRIC INDUSTRY

While legislation establishes the basic policies for electric power, the implementing of these policies is in the hands of the various regulatory agencies. While all of the previously cited agencies play a role in the regulation of the electric power industry, FERC plays the most important role. FERC is in the process of implementation of the "'92 EPACT". It has done so with a series of orders; Orders 888 and 889 and Order 2000. Presently it is continuing its activities with the SMD-NOPR. A review of these key FERC decisions and procedures is essential to understanding the functioning of the electric power industry. The implementation steps FERC mandated have involved a multi-stage approach as lessons have been learned as the process has evolved.

Orders 888 and 889

FERC Issued Orders No. 888 and No. 889 in April 1996, intending "to remove impediments to competition in the wholesale bulk power marketplace and to bring more efficient, lower cost power to the Nation's electricity consumers." Further, they were to "... remedy undue discrimination in access to the monopoly owned transmission wires that control whether and to whom electricity can be transported in interstate commerce."

These orders had three goals:

1. Promotion of wholesale competition through open access, non-discriminatory transmission services by public utilities;
2. Recovery by public utilities and transmitting utilities of stranded costs;
3. Establishment of an open access same time information system and standards of conduct.

All public utilities that owned, control, or operate facilities used for transmitting electric energy in interstate commerce were required:

- To file open access non-discriminatory transmission tariffs that contain minimum terms and conditions of non-discriminatory service[15];
- To take transmissions service (including ancillary services) for their own new wholesale sales and purchases of electric energy under the open access tariffs;
- To develop and maintain a same time information system that will give existing and potential transmission users the same access to transmission information that the public utility enjoys (OASIS).
- To separate transmission from wholesale power service functions (functional unbundling instead of the more draconian step of corporate unbundling) meaning that they:
  - Take wholesale transmission services under the same tariff of general applicability as they offer their customers;
  - State separate rates for wholesale generations, transmission, and ancillary services;
  - Rely on the same electronic information network that their transmission customers rely on to obtain information about the utilities' transmission systems.

The orders also:

- Clarified federal/state jurisdiction over transmission in interstate commerce and local distribution;
- Identified ancillary services[16] required for the functioning of the transmission system;
- Provided guidance regarding the formation of Independent System Operators (ISOs) which would be subject to FERC's authority, which will be discussed in Chapter 13.

[15] The Order included a pro-forma open access tariff.

[16] *See* Chapter 8 for a list of ancillary services.

Eleven principles for ISOs were identified to ensure that the operation of the transmission system would not favor the utility owners. ISOs must:

1. Be structured in a fair and non-discriminatory manner;
2. Have no financial interests in any market participant (neither the ISO nor its employees);
3. Provide open access at non-pancaked rates, that is, a single unbundled grid-wide tariff;
4. Have primary responsibility for the short-term reliability of the grid—by complying with NERC and regional standards;
5. Have control over the operation of the interconnected transmission facilities;
6. Identify constraints on the transmission system and take operational actions to relieve them;
7. Have pricing policies for transmission and ancillary services that promote efficient use of investments in generation and transmission;
8. Have incentives for efficient management and administration procured in the market;
9. Develop mechanisms to coordinate with their neighbors;
10. Establish an alternate dispute resolution process;
11. Make transmission system information publicly available (OASIS).

Order 2000

In December 1999, FERC issued Order 2000 dealing with Regional Transmission Organizations (RTOs). FERC stated "continued discrimination in the provision of transmission services by vertically integrated utilities may also be impeding full competitive electricity markets" and that its "goal is to promote efficiency in wholesale electricity markets and to ensure that electricity consumers pay the lowest price possible for reliable service." To address this concern, FERC took its next step in defining its requirements for an independent transmission provider. It required ". . . that each public utility that owns, operates, or controls facilities with the transmission of electric energy in interstate commerce make certain filings with respect to forming and participating in an RTO."

- It established minimum characteristics and functions for RTOs.
- It stated it would sponsor a collaborative process by which public utilities and non-public utilities that own, operate, or control interstate transmission facilities may form or join RTOs; in consultation with state officials as appropriate.
- It gave a time line for public utilities to make filings to initiate operation of RTOs.

FERC specified four minimum characteristics and eight functions of an RTO.[17] To some extent these mirrored the eleven principles specified for ISOs. They differed in that the RTO material included provisions for planning and expansion and for management of congestion and parallel path flow.

Tariff Basis

A key FERC requirement is that transmission systems must be open to all potential users on a non-discriminatory basis. This involves the rights to use the facilities and the tariffs to be charged. It requires that all others can also use a transmission system built to supply a specific group of consumers on the same basis as the original users.

A major item of contention is the return that should be allowed on new transmission investments, with transmission owners and potential transmission investors considering the rates of return on investment proposed by FERC to be too low to provide needed profits and to justify the significant additional investment that is required.

Transmission Rights

Transmission rights define property rights and not only entitle market participants to the financial benefits associated with the use of a transmission system, but also encourage investment in transmission infrastructure by providing a commodity that is fixed in nature and can be traded in an open market.

There are two types of transmission rights: physical transmission rights (PTRs) and financial transmission rights (FTRs).

Physical Transmission Rights. PTRs provide the owner with the exclusive right to transport a pre-defined quantity of power between two locations on a transmission network and, importantly, to deny access to the network by market participants that do not hold transmission rights. Further, PTRs offer the two necessary features of transmission rights. They provide clearly defined property rights because once the PTR is purchased, the holder of that right is assured that capacity will be reserve exclusively for the transmission of power under that right. Also, during periods of high demand, the PTR can be sold to another party giving them the right to use that same capacity. With a PTR, transmission costs can be determined in advance.

PTRs present numerous problems. First, the right of the PTR holder to self-dispatch can interfere with the transmission system operator's efforts to effi-

[17] The minimum characteristics of an RTO were specified as: independence, scope and regional configuration, operational authority, and short-term reliability. The minimum functions of an RTO are: tariff administration and design, congestion management, parallel path flow management, ancillary services, OASIS mode and calculate and publish on it total transmission capability (TTC) and available transmission capacity (ATC), market monitoring, planning and expansion of the transmission system, interregional coordination of both reliability and market issues.

ciently schedule and dispatch the system. This can interfere with the reliability of the operations of the transmission system. This reliability issue can be addressed by issuing an amount of PTRs that represent less than the full capacity of the system. However, this is inefficient and would prevent cost savings from being realized that result from increasing output in inexpensive locations and decreasing output in more expensive locations.

Moreover, the PTR holder's ability to exclude access to the transmission market presents a problem. If a generator holds a PTR for transmission service from point A to point B and owns generation at point B, it could maintain an artificially high price at point B by preventing generators at point A from transmitting power across the transmission system. Thus, the most efficient generators might be at point A, but by withholding transmission capacity, the generator at point B could maintain pricing inefficiencies for its gain.

Financial Transmission Rights. FTRs are contracts between a market participant and the transmission system operator. While FTRs are similar to PTRs in that they are defined from a source to a destination and are denominated in a MW amount, the key difference is that FTRs do not entitle their holder to an exclusive right to use the transmission system. Instead, FTRs are tradable rights that are automatically assigned to those users that provide the system with the highest value. Importantly, FTRs have no adverse impact upon economic dispatch because the contracts underlying the FTRs are settled outside of the physical dispatch of the system.

FTRs provide their holders with the right to payments equal to the energy price difference between the source and the destination locations. These payments are funded by the congestion payments that arise when energy is purchased from lower-priced regions to be sold in higher-priced regions and there is insufficient transmission capability causing congestion on the transmission system.

FTRs remain one of the most dynamic of issues, since it requires an entirely new perspective at how to fund grid expansion. These issues are complex. FTRs can be viewed as *physical*, offering effectively an admission ticket to allow the holder to schedule a transmission transaction, or as *financial*, as favored in the Northeast, which offers the holder a stream of revenues even if it does not deliver or receive physical energy. It has become apparent, however, that the level of congestion payments is insufficient to induce construction of new transmission to relieve transmission bottlenecks.

Average System versus Incremental Costs

Another major question is the cost to be charged for use of a transmission system. For example, when a generator is added to the system, a number of transmission costs may be involved:

1. The costs to connect the generation to the system;
2. Reinforcement of or additions to the transmission system, either in the vicinity of the generation addition or possibly hundreds of miles away in other systems;
3. Changes and replacements in protection and control systems, such as circuit breaker replacements, and new relaying arrangements;
4. The increase in costs resulting from transmission outages while the necessary connection and reinforcement work is done.

One approach is to have the new generator pay all of the above additional costs. Additional transmission costs can also be incurred if a generator unit is retired, causing increased loading on certain transmission facilities. This raises the question of when the costs would otherwise be incurred if the generator addition or retirement were not made.

The other approach is to include all transmission costs except those of the direct connection in overall system tariffs. This raises the question of why should the existing generators and consumers have to pay more for transmission service because one party has added or retired a generator? It also raises doubts about the selection of the most economic generation additions since the system costs, which are subsidized by all customers, may be much larger for one than another, but do not enter into the competitive decision process.

STATE REGULATION

Complicating the development of the regional markets were differing approaches that individual states took with respect to utility restructuring. State activities with respect the utility restructuring tended to mirror the relative cost of electricity in their states. States with higher costs generally implemented more aggressive restructuring, over and about what FERC required, than those with lower costs.

The experience in California has severely reduced the popularity of state restructuring activities. State restructuring activities included:

- Some form of immediate rate reduction and freezing for a specified period thereafter with provision for stranded cost recover;
- Mandated or encouraged generation plant divestiture;
- Requirements for retail access.

Customer Choice

A major advantage claimed for restructuring is the provision of "customer choice." Customers would have the ability to decide from which companies

Status of State Electric Industry Restructuring
Activity As of February 2003

Retail access is either currently available to all or some customers or will soon be available:

- ❑ Arizona, Connecticut, Delaware, District of Columbia, Illinois, Maine, Maryland, Massachusetts, Michigan, New Hampshire, New Jersey, New York, Ohio, Oregon, Pennsylvania, Rhode Island, Texas, and Virginia. (In Oregon no customers are currently participating in the State's retail access program, but the law allows non-residential customers access).

States are not actively pursuing restructuring:

- ❑ Alabama, Alaska, Colorado, Florida, Georgia, Hawaii, Idaho, Indiana, Iowa, Kansas, Kentucky, Louisiana, Minnesota, Mississippi, Missouri, Nebraska, North Carolina, North Dakota, South Carolina, South Dakota, Tennessee, Utah, Vermont, Washington, West Virginia, Wisconsin, and Wyoming. (In West Virginia, the Legislature and Governor have not approved the Public Service Commission's restructuring plan, authorized by HB 4277. The Legislature has not passed a resolution resolving the tax issues of the PSC's plan, and no activity has occurred since early in 2001).

States where there is a delay in the restructuring process or the implementation of retail access:

- ❑ Arkansas, Montana, Nevada, New Mexico, and Oklahoma.

California is the only state where direct retail access has been suspended.

Figure 11.2. Status of state electric industry restructuring activity as of February 2003.[§§§§]
[§§§§] Information extracted from information at EIA Website: *www.eia.doe.gov*

would provide their electric power. Theoretically, this competition might lead to lower electricity prices. The existing distribution system would continue to deliver the power, and would continue to be regulated. Figure 11.2 shows the states that have enabled customer choice as of February 2003. In general, a very small percentage of the customers having the ability to choose another power supplier have used it. Of those that have switched the majority have been industrial and large commercial customers.

Metering

Customer choice has required significant additional metering expenses so that the data required for the billing of supplies to many consumers having different suppliers could be handled. It has also become necessary to meter use over precise time frames to accommodate new usage rates.

Distribution Rates

In the past, distribution system rates were based on imbedded investment costs and actual operating costs. This procedure is being continued. However, increased attention is being placed on performance based rates, that is, rates that depend on the quality and reliability of service, and the overall response to customer needs.

State and Local Environmental Requirements

State and local laws cover a number of environmental and safety matters. Local communities often must issue building permits. These requirements can make very difficult the ability to build transmission and transmission facilities needed in a regional basis. Under various Federal Environmental Laws, state authorities have been given responsibility for ensuring utility compliance. Insofar as transmission line approvals, almost all states require examination of environmental compatibility, as well as public need prior to approving new transmission lines.

OVERALL REGULATORY PROBLEMS

The legislative and regulatory networks discussed above are obviously complex and often create problems because of:

- Lack of understanding by regulators and legislators of the technical and economic functioning of electric power systems;
- Conflicts between states and the federal government over policy and authority;
- Conflicting regulatory priorities, that is, environmental concerns versus the requirement of low priced electricity.

Many believe we need far greater coordination between our legislation, regulation and power system technology.

12

THE BUSINESS NETWORK

INVESTMENT AND COST RECOVERY

As with any other business, the investors in the electric power business expect to receive a return on their investment commensurate with the risks involved and a return of their investment over a reasonable period of time. Prices must be set to enable the investors to recover all their costs, including taxes, plus a return on their investment, and a return of their investment. The return on their investment, they believe, should justify the perceived risks they are taking. Ultimately, the economic justification for investments depends on the resulting costs for the product or service. While prices must be adequate, they are affected by commercial policies, subsidies, taxes, and other factors which are subject to change and man-made decisions.

The electric utility business from its inception has been capital intensive with the expensive facilities installed having long service lives and, therefore, long periods required for the recovering of the investments involved.

Both engineers and economists had found in the past that a vertically integrated power company having exclusive rights to serve a "franchised" area provided significant benefits since it facilitated the long-range optimization of the total supply system from the fuel source to the customer's meter. Significant cost reductions also resulted from a centralized planning, operating, purchasing, legal and administrative staff.

Important financial advantages were also possible since revenue from one portion of the business would help to provide funds if a large unanticipated

Understanding Electric Power Systems: An Overview of the Technology and the Marketplace, by Jack Casazza and Frank Delea
ISBN 0-471-44652-1

cost increase occurred in another portion of the business. An example is the need for large amounts of money to repair a distribution system after a major hurricane. With a vertically integrated organization funds from power sales and transmission system use could help provide the required funds. This diversification of risk was an important asset of the vertically integrated utility business. However, some believed it prevented competition, competition which could produce significant benefits.

CHANGING INDUSTRY STRUCTURE

In the 1980s and 1990s a significant movement developed in the United States for a change in the structure of electricity supply industry. Supporters of restructuring proposed that the vertically organized electric utility, with captive customers and regulated by government, be replaced with a more competitive environment. For competition in the sale of any commodity to be effective, all potential suppliers must have the ability to deliver their product to the customers.

It was not always recognized that key to the success of competition is that ways must be developed to reward those who improve productivity and who create additional savings. Because of the unique nature of electric power systems, major difficulties arose in achieving this result. Many believe the changing industry structure has increased costs and decreased reliability, and is merely a means of re-allocating costs and profits.

Utility Responses

In addition to the actions mandated by FERC, utilities responded to restructuring pressures and state requirements in a variety of other ways by:

- Establishing themselves as subsidiaries of newly formed holding companies;
- Entering the EWG market, both in the United States and overseas, through unregulated subsidiaries;
- Divesting their power plants;
- Seeking mergers to acquire economies of scale as well as deeper pockets to engage in non-regulated activities.

Holding Company Formation. Reversing the reduction of holding companies that the SEC encouraged in the 1935 to early 1950s period, utilities sought to establish holding companies to give them the flexibility they felt they needed to survive in the new environment.

Figure 12.1 lists 28 registered holding companies in existence as of October 2002. About one-third of these companies had subsidiaries which also were holding companies. It is common to find a number of regulated and unregu-

Registered Holding Company Name	Electric System (E) Gas System (G)	Number of Subsidiary Holding Company
AGL Resources, Inc.	G	0
Allegheny Energy, Inc.	E&G	1
Alliant Energy Corp.	E&G	0
Ameren Corp.	E&G	0
American Electric Power Co., Inc.	E&G	1
CenterPoint Energy, Inc.	E	0
Cinergy Corp.	E&G	1
Dominion Resources, Inc.	E&G	1
E.On ag	E&G	12
Emera, Inc.	E	2
Energy East Corp.	E&G	0
Entergy Corp.	E&G	0
Exelon Corp.	E	4
FirstEnergy Corp.	E&G	0
Great Plains Energy, Inc.	E	0
KeySpan Corp.	E&G	0
National Fuel Gas Co.	G	0
National Grid Group plc	E	11
Nisource, Inc.	E&G	1
Northeast Utilities	E&G	0
Pepco Holdings, Inc.	E&G	3
Progress Energy, Inc.	E&T	0
SNACA Corp.	E&G	0
Scottish Power plc	E&G	0
Southern Co.	E	0
Unitil Corp	E&G	0
WGL Holdings, Inc.	G	0
Xcel Energy, Inc.	E&G	0
TOTAL:	28	37

Figure 12.1. Holding companies registered under PUHCA as of October 31, 2002.*****
***** Information extracted from table at SEC Website: *www.sec.gov/divisions/investment/regucacompanies.htm.*

lated subsidiaries under the umbrella of a parent holding company. In some instances, the non-regulated subsidiaries became so large they were spun-off into completely separate companies.[1]

Unbundling. Economic theory requires effective competitive markets be established to achieve the benefits of competition. The first step selected in this process was the requirement for utilities to "unbundle", that is, make separate businesses out of their generation, transmission, and distribution systems. At first, this required keeping separate books of account for each business. Subsequently pressure increased to make separate companies for each system.[2]

[1] For example, the Southern Company spun off its non-regulated generation subsidiary into a new company Mirant.
[2] The first system to form separate companies was the Oglethorpe Power Company, which unbundled in 1997 to form a generation company, and transmission company, and a system operation company.

New Structure

Where full unbundling occurred, the participants in the new industry structure became:

- Power producers;
- Transmitters;
- Distributors;
- Power marketers.

Power Producers. The power producers include:

- The existing investor owned, cooperative, municipal and government owned systems who retained ownership of their existing generators and, in some cases, installed additional generation.
- Non-utility power producers who either purchased existing power plants or installed new ones to produce power for sale to others.
- Federal systems producing power for sale, including TVA, BPA, SWPA, and so forth.

The generating capacity of each of these groups is shown in Figure 1.4.

Power Plant Divestitures. In four states,[3] state law required power plant divestiture. In others, divestiture was tied to the utilities' efforts to secure recovery of stranded investment. The market value of a power plant was considered to be the surest way to determine the stranded cost of a power plant.

Divestitures were of two types:

1. To a non-regulated subsidiary of the utility;[4]
2. To an independent, non-regulated company.

In most cases, the independent companies purchasing the generation assets were non-regulated subsidiaries of holding companies which had also a regulated company subsidiary.[5] However, there were also purchasers who were part of the new generation of non-regulated power companies.[6]

Because of concerns about the purchaser acquiring market power, if a single buyer acquired all of a utility's generation assets, many states limited the amount of capacity that any one entity could buy.

[3] California, Connecticut, Maine, and New Hampshire. Texas placed a limit on the amount of generating capacity any one entity could have.

[4] Illinois Power, Public Service Electric & Gas.

[5] US Generating—PG&E, Houston Industries/Reliant Power—Houston L&P, FPL Group—Florida P&L, Edison Mission Energy—SoCal Edison.

[6] Calpine Energy, Sithe, Dynergy.

Plant Name	Old Licensee	New Licensee	Transfer Approval	Capability Net mWE
TMI-1	GPU, Inc.	AmerGen	April 12, 1999	786
Pilgrim	Boston Edison Co.	Entergy	April 29, 1999	655
Clinton	Illinois Power Co.	AmerGen	Nov. 24, 1999	930
Vermont Yankee	Vermont Yankee	AmerGen	July 7, 2000	506
Oyster Creek	GPU, Inc.	AmerGen	Aug. 8, 2000	619
Fitzpatrick	New York Power Authority	Entergy	Nov. 9, 2000	820
Indian Point 3	New York Power Authority	Entergy	Nov. 9, 2000	970
Millstone 1	Northeast Utilities	Dominion	Mar. 9, 2001	Note 1
Millstone 2	Northeast Utilities	Dominion	Mar. 9, 2001	872
Millstone 3	Northeast Utilities	Dominion	Mar. 9, 2001	1,146
Nine Mile Point 1	Niagara Mohawk	Constellation	June 22, 2001	619
Nine Mile Point 2	Niagara Mohawk	Constellation	June 22, 2001	1,154
Indian Point 2	ConEdison of NY	Entergy	Aug. 27, 2001	953

Last revised Friday, December 06, 2002 www.nrc.gov/reading-rm/doc-collections/fact-sheets/transfer.html.
NOTE 1: Millstone 1 was permanently closed November 1995.

Figure 12.2. Fact sheet on NRC reactor license transfer.

NRG, Reliant Industries, U.S. Generating Company, Mission Energy and the Southern Company were among the companies purchasing the largest amount of fossil/hydroelectric capacity. Based on published information, purchase prices for hydroelectric and fossil fuel plants were usually in excess of book value—sometimes well in excess. Subsequent to the collapse of ENRON many merchant energy providers were forced to sell power plant assets they had purchased only a few years earlier. Unlike the earlier transactions, these sales were well below the purchase price.

There are two main explanations given of why companies were willing to pay more than book value. The purchasers thought that:

1. They would be able to recoup the cost of the plants in the higher energy prices in the new markets being formed.
2. The sites that the plants occupied had an intrinsic value because of the availability of transmission, water, and fuel delivery systems.

Perhaps the most surprising occurrence was that there was a market for nuclear power plants. Figure 12.2 lists those nuclear plant's license transfers that the Nuclear Regulatory Commission approved. The figure does not give a complete picture, since minority ownership interests were separately purchased.[7]

As the figure shows, Entergy Nuclear, Dominion Resources, and Mission

[7] Two small owners of Seabrook sold their interests to Great Bay Power (2.9%) and Select Energy (3 mW), Duquesne Light swapped its ownership interest (300 mW) in the Beaver Valley Units for fossil units from FirstEnergy, Conective sold its minority ownership interest in Hope Creek, Peach Bottom, and Salem Harbor to PSE&G Power and PECO.

Energy were the biggest purchasers. The new owners believe that their skill in operating nuclear plants, rising fossil fuel prices, and increasing air pollution concerns will make their investment profitable.

Experience has shown that forced divestiture will not solve market power abuses. In fact, forced divestiture can create market power problems especially if there is a shortage of generating capacity in the region. To protect consumers from possible market power abuses, California's shareholder-owned electric companies were "strongly encouraged" by the state's restructuring law to sell their fossil fuel-powered generating assets. Electric companies in California were required to buy power on the "spot market." Long-term, fixed-price contracts were not permitted, preventing utilities from locking in stable prices over a long period. The sale of generating assets meant that the distribution utilities did not own enough capacity to meet their customers' demands. When spot market prices for electricity spiked, shortages, rolling blackouts, and financial chaos resulted from electricity providers and consumers.

Forced divestiture was but one of the many factors that helped bring about the recent breakdown of the California electricity market. But forced divestiture played a key role by denying companies the needed capacity to fully serve their customers' needs.

Transmitters. Originally, transmission was provided by the vertically integrated utilities and Federal agencies such as TVA and Bonneville. These systems later interconnected with each other to achieve economic benefits as discussed in Chapter 6, which also discusses the functions of the transmission system. With the advent of restructuring, the purpose of the transmission system was reoriented to become the means of facilitating competition in the bulk power electric market in the sales and purchases of electric energy.

This new requirement diminished the attention given to the other transmission functions previously discussed, particularly those related to reliability. The transmission system role in reducing total generating capacity requirements[8] often became neglected and transmission constraints became more and more frequent.

The United States transmission systems are aging, with some important facilities having ages in excess of seventy five years. Very little in the way of new transmission facilities have been added over the last ten years because of uncertainties about the revenue that could be earned by the transmission business and the concern about their use to deliver power supplies from other sources, reducing the earnings from an existing generation business. Meanwhile, competitive pressures to reduce costs caused reductions in maintenance expenditure.

The combination of these effects has lead to an aging transmission system with increasing outages and a large backlog of maintenance requirements. This has lead to the FERC requirement to form regional transmission organiza-

[8] This is often called the capacity benefit margin or CBM.

tions to operate and/or to plan transmission systems, specifically Regional Transmission Organizations (RTOs), Independent System Operators (ISOs), and Integrated Transmission Providers (ITPs), all of which are discussed in the next chapter.

At present, increasing attention is also being given to "merchant transmission" companies that would make investments in specific transmission lines and be directly compensated for their use.

Development of Non-Regulated Power Market

As a result of power plant divestiture and the construction of new primarily gas-fired power plants by EWGs, the amount of generation supplied by utility and non-utility generators has changed significantly over the 1990s, as shown on Figure 12.3.

In the decade following EPAct, the number of entities allowed by FERC to charge market based rates has exploded. There are well over 1,400 such entities as of early 2003.

Distributors

The final delivery of electric power is accomplished by the subtransmission and distribution systems except for a few very large consumers who are supplied by the transmission system. With unbundling, the distribution system is paid only for the delivery service it provides, with its charges determined based on traditional cost of service approaches and approved by the state regulators.

Period	Industry	Utilities	Non-Utilities
1991	3,071,201	2,825,023	246,178
1992	3,083,367	2,797,219	286,148
1993	3,196,924	2,882,525	314,399
1994	3,253,799	2,910,712	343,087
1995	3,357,837	2,994,529	363,308
1996	3,446,994	3,077,442	369,552
1997	3,494,223	3,122,523	371,700
1998	3,617,873	3,212,171	405,702
1999	3,704,544	3,173,674	530,871
2000	3,799,944	3,015,383	784,561

NOTES: Values for the industry and non-utilities for 2000 are preliminary; utility values for 2000 are final. Values for 1999 and prior years are final. Due to restructuring of the electric power industry, electric utilities are selling plants to the non-utility sector. This will affect comparisons of current and historical data. DATAs may not equal sum of components because of independent rounding.

SOURCES: Energy Information Administration, Form EIA-759, "Monthly Utility Power Plant Report," Form EIA-900, "Monthly Non-Utility Power Plant Report," and Form EIA-860B, "Annual Electric Generator Report Non-Utility" (and predecessor forms).

Figure 12.3. Net generation, 1991 through 2000 (*Million Kilowatt-hours*).

The distribution system not only delivers electric power, but operates as an important element in the functioning of the overall system. It participates in the provision of system reactive supply and voltage control. It has underfrequency relaying installed to shed load and limit major bulk supply system disturbances when system frequency decline.

Marketers

With unbundling and restructuring of the bulk power market it became feasible for electric power suppliers and large consumers to choose from a number of power suppliers and a number of transmitters. This offered an opportunity to form marketing organizations to buy and sell power, trading sometimes a day or two ahead of time and sometimes hourly.

These marketing organizations were sometimes in the corporate structure of existing large utilities also having considerable generation and transmission assets and sometimes purely trading organizations.[9] Theoretically, these trading organizations would create economic benefits by enabling purchasers to obtain lower cost power and, through competition, driving down overall power costs. These organizations also provided a means of guaranteed future electricity costs, hedging against potential volatile prices. This was a form of insurance for which the trading company could earn revenue, but which increased the cost of electricity.

Practically, many of the trading organizations became a mechanism for "gaming" the market to increase their profits at considerable expense to the overall American public. This "gaming" was accomplished in some cases through fraud and false bookkeeping and accounting. In other cases it was accomplished through legal manipulation[10] of the systems and schedules, and collusion with the power providers.

At first, the growth in activity by marketers was phenomenal, but recent criminal and other investigations is resulting in the demise of most marketing organizations. Banks and reputable financial institutions seem destined to be the surviving marketing organizations.

Wheeling and Customer Choice

Wheeling is a term sometimes used to describe the transmission of power by one system for another system. Wheeling is an essential ingredient in the move to provide customers with the choice of power supplier.

A significant number of states provide customers with the ability to select their power supplier as was shown in Figure 11.2. In cases where the customer

[9] Enron was this type of organization.

[10] While legal, some consider these manipulations a violation of the code of ethics for the professions involved.

selects a supplier other than their local utility, some form of wheeling service is needed along with local delivery service.

In cases where an alternate supplier is chosen, the local utility rates are reduced by the revenue requirements associated with the capital and energy costs of the generation it is no longer supplying, leaving only distribution delivery costs. These "offset" costs are those that an alternate supplier must beat to result in a lower total cost to consumers. Experience to date has shown relatively little customer switching since alternate suppliers generally have higher power costs than the present supplier.

In the early stages of restructuring, many companies attempted to develop businesses supplying customers in other systems. Many of these efforts were not successful, often resulting in economic losses. As a result, the level of competition in the supply of retail customers has decreased significantly.

Contracts and Agreements

With an unbundling electric utility industry, many new contracts are required, including contracts for power purchases and contracts for transmission services. With the vertically integrated utility, these requirements were coordinated under the direction of a single corporate management. With restructuring, the terms and conditions of the required services must be covered by appropriate contracts.

In addition, "coordination contracts" can help to achieve lower total system costs. Because actions by any one company, or failure to act, can affect the costs of other companies agreements for sharing the costs and benefits of coordination. Such coordination contracts are becoming increasingly necessary. One example would be installation of transmission facilities in one system to provide for reduced costs in another system.

13

ISOs, RTOs AND ITPs

In response to the requirements of Orders 888 and 889, which are discussed in Chapter 11, the utility industry embarked on a major voyage of change which continues to the date of writing this book. The most noticeable of the immediate actions were the filing by all transmission providers with FERC of new open access tariffs to be used to pay for use of transmission systems, the separation within companies of the generation/marketing function from the transmission function and the efforts by various operating entities to develop real time systems to report on use and capability of the transmission system in their areas.

FERC allowed some flexibility in how the private sector utilities functionally separated (unbundled) their transmission functions from their generation and marketing functions. Some utilities simply separated within the same corporate structure by imposing restrictions on contacts between individuals working in the respective areas and by separating the respective costs within their accounting system. Others went a step further and formed ISOs.

ISO FORMATION

A number of ISOs were formed to manage the transmission grid.[1] These new ISOs were primarily the existing power pools with their governance provisions

[1] As of mid-2002, there were five ISOs: New England, PJM, New York, California, and ERCOT.

Understanding Electric Power Systems: An Overview of the Technology and the Marketplace, by Jack Casazza and Frank Delea
ISBN 0-471-44652-1

changed to separate the operation and control of the transmission system from its owners and to have governance provisions acceptable to FERC. The formalized rules ensuring this separation were a central issue in FERC's process to approve each of the ISOs.

The issue of the wholesale market structure was also handled in different ways. Some proposed a completely separate entity which would operate/administer the wholesale energy market. Others felt that the function should be a responsibility of the ISO.

Each region was allowed to formulate its own rules for the operation of the local power market. It was FERC's stated intent not to specify how the power markets were to be structured. FERC's focus was on the forms of the organizational structure of the market's management to ensure openness and not on the mechanics of how the market was going to work. The actions taken to develop the power markets reflected the lack of specific guidelines within FERC's order, resulting in many different market structures. Even those markets supposedly using the same approach, that is, locational marginal pricing, differed in their details.

The different market structures had an impact on the resolution of the seams issues. There are a number of definitions of seams issues, the following are our own:

- Seams issues refer to issues which arise when there are power transfer conditions which involve more than one area, each of which uses different operational and/or market rules. These differences in rules can adversely impact the overall economic performance of the electric power system in two ways:
- Economic transactions between adjacent areas or through an intermediate area could be hindered or prevented because of differing market rules.
- Reduction of power flows to ensure that power transfer limits are observed may not be done in a manner to have the lowest impact on the price of electricity.

Functions of ISOs

Regional Operating Functions. In Order 888, eight operational functions required of each ISO were included in the Order's eleven principles. As with the wholesale energy market, areas had latitude on how to address the principles. The specific approaches to transmission operation differed from region-to-region, but the areas addressed were similar:

Transmission System:

- Calculation of available transmission capability and making these values known by posting on the OASIS system;
- Scheduling transmission and generation maintenance outages in a manner that minimizes the impact on the operation of the grid;

- Having either hands-on or contractual control of the operation of the transmission system;
- Ensuring that adequate ancillary services are available for the reliable operation of the transmission system.

Additionally, the ISOs that were formed used a form of bidding for their energy markets. The purpose of the wholesale market was:

- Manage a longer term (day ahead) or a shorter term (hour ahead) market in which capacity, energy, and ancillary services are based on
  - Bilateral contracts and bids received via the OASIS system;
  - Available transmission capability.

Regional Planning Functions. The ISO also has the responsibility for ensuring that adequate resources are provided. In some ISOs generation adequacy is provided by specifying an "Installed Capacity Requirement" (ICAP). A part of this responsibility includes the forecasting of future peaks and seasonal loads and comparing expected generating capacity a few years in the future. If adequate generating capacity is not expected to be available, the ISO takes steps to make this known. Appropriate action would be required of the regional reliability council, NERC, and FERC to correct the deficiency. This includes those short generating capacity being required to make capacity payments commensurate with their shortage.

In addition, the ISO has the responsibility of reviewing and/or developing long-range transmission plans for the region. This is a difficult and uncertain process since future generation additions will be determined in the competitive generation market and are difficult to estimate. A major concern is that transmission additions will be based either on meeting only immediate needs or on long-range generation developments that may not materialize. In either event, there is high risk of considerable extra transmission costs or an inadequate transmission systems.

RTOs

With Order 2000, FERC specified in more detail its requirements for independent transmission operations. However, it still did not specify the market structure or the size that an RTO must take to meet the minimum characteristics and functions.[2] "The characteristics and functions could be satisfied by different organizational forms, such as ISOs, Transco's, combinations of the two, or even new organizational forms not yet discussed in the industry or proposed to the Commission. Perhaps anticipating a continuing evolution in its

[2] FERC Requirements for RTOs were listed in Chapter 11.

requirements, FERC also indicated: "We also establish an 'open architecture' policy regarding RTOs, whereby all RTO proposals must allow the RTO and its members the flexibility to improve their organizations in the future in terms of structure, operations, market support, and geographic scope to meet market needs."

The Order did not address the economic impact of different market structures. FERC seemed also to be relying on having very large RTOs as a way of eliminating seams issues.

The Order was not uniformly accepted across the country. There were two main objections especially relating to the multi-state size of the RTOs that was envisioned:

- State regulators were concerned about their apparent loss of control over the utilities supplying customers in their states.
- States in the South and the Pacific Northwest, currently experiencing lower-cost electricity, were concerned that the development of a national transmission grid would impact their cost of power because their cheap power would be diverted to higher cost regions.

The RTOs would also be "not-for-profit" companies. As part of FERC's evolving position, a debate continues, however, on whether the operation of the transmission system should be on a "non-profit" or "profit making" basis as will be discussed in Chapter 14.

The utility response to Order 2000 varied region by region. Even within regions, as the difficulty of reconciling different operating and market rules to form large RTOs became apparent, utilities shifted from one combination of companies in one structure to different combinations.

In the eastern region of the United States three of the ISOs, PJM, ISO New England, and the ISO New York, embraced the "PJM style" market design for their proposed RTOs. The biggest issues bearing on consolidation of the eastern ISOs into a super-RTO are loss of control by market participants and regulators over the existing ISOs, and potential cost shifting from one subregion to another if the markets are merged.

The southeastern states have resisted RTO/ISO formation for quite a while and want a model that would maintain the status quo as much as possible. This region has been working diligently to develop an RTO that would be acceptable to state and federal regulatory bodies.

Midwestern states are moving forward with some entities favoring either the MISO or PJM. The west has been unable to form a single broad RTO and common market. The result will be the formation of two new entities: RTO West in the Pacific Northwest, and WestConnect, covering the Southwest.

Texas is an "island" and wishes to remain that way. It is unique in that both wholesale and retail are under the jurisdiction of a central administrative entity.

14

THE MONEY NETWORK

A vital part of the electric power business is the functioning of its money network. This network consists of the various sources of money, the various paths and organizations through which this money flows, and the money used for various reasons, including payments of bills, debts, and profits.

This money network is tied to the physical and energy networks at various points, with the flow of money having important effects on their development and operation. The money network is controlled by both the business and regulatory networks, and uses the generation and communication networks.

Money sources are similar to generators in electric power systems. These sources are the investors, consumers, taxpayers, banks, and so forth. By drawing the diagram for the money network and analyzing the sources of money, its flow through the networks, and the uses, charges, or profits for various functions, we can get a far better understanding of how our financial systems work and are related to electric power policy. These money networks have been diagramed in the past by engineers, starting in the early 1960s.[1] It is important to identify the connection points between the money network, the energy network, and the power network. What happens to one affects the others. (Analyses of this network would have provided an early detection of Enron's procedures. It would also show what happened to the $38 billion in extra costs that occurred in California.)

[1] See CIGRE paper in 1962, by Sels and Dillard.

Understanding Electric Power Systems: An Overview of the Technology and the Marketplace, by Jack Casazza and Frank Delea
ISBN 0-471-44652-1

ALLOCATION OF COSTS AND ECONOMIC BENEFITS

In the past, the tariffs paid by all consumers from large industries to small residential consumers were established by the states. Theoretically these tariffs were designed to reflect the costs to supply each class of customer. Practically, political considerations often overrode a fair cost allocation, with large industry frequently charged more than its share to subsidize small consumers, who were a majority of the voters. Large industry often felt its competitive positions were adversely affected by regulatory decisions, and their answer was to get rid of regulation.

The participants in joint power projects and coordination procedures allocated costs between the participants based on negotiations between the participants, with regulatory oversight and approval. Prior to the restructuring of the electric power industry, coordination between the supplying systems took place to achieve economic benefits. To allocate costs and benefits in projects involving more than one system, a "split savings" approach was used. This approach was based on review of costs to each of the participants without the joint project and the costs with the project. The benefits from the coordinated procedures were allocated to each participant, sometimes 50/50 and sometimes based on negotiations. This approach was used for projects involving capital investments and for operations such as generation dispatch. It was a key factor in determining the flow of money.

If extra costs were required of any participant, they would be compensated for these in addition to receiving their share of the net benefits. With this procedure, all participants in coordinated activities would benefit. It would be a "win-win" situation for all. This procedure produced a great deal of cooperation and large savings to the public, estimated to be more than $20 billion per year in the late 1980s.[2] These costs and associated benefits were then reflected in the tariffs paid by consumers.

New restructuring procedures rely increasingly on market forces and competition to achieve cost allocation, especially for generation and the power the plants produce. Major questions exist, however, about the ability of some organizations to exert "market power" dominating the market, and controlling pricing. The smaller organizations are concerned with the ability of the large ones to drive them out of business.

With the advent of restructuring the coordination achieved with the split savings approach has ceased. The revised approach is the use of market forces and competition to spur cost and price reductions and more economic solutions. The revised approaches are a key factor in determining the money flow among the participants. Results to date indicate restructuring has increased overall natural costs for electricity by about 10%[3] from what they would otherwise have been.

[2] Casazza, J.A., Palermo, P.J., Lucas, J., and Branco, F., *Generation Planning and Transmission Systems*, CIGRE Paper. Also EEI publications.

[3] "Electricity Choice: Pick Your Poison—Errant Economics? Lousy Law? Market Manipulation? All Three!", J.A. Casazza, Public Utilities Fortnightly, March 1, 2001.

Average Costs Versus Incremental Costs

An additional concern in the operation of the new competitive markets is whether prices should be based on average costs or incremental costs. The average cost approach is simpler and often less contentious. The incremental cost approach is often fairer since each use or user of the system will pay for the costs it causes. This question becomes important in determining tariffs and the allocation of additional investments.

MARKET VERSUS OPERATIONAL CONTROL

In an electric power system, there are two arrangements for the control of decisions. Those made by those having "market rights" developed through operation of the money network, and those having the operational control of the physical network.

Theoretically, the operation of both of these networks should be coordinated. The operation of the money network should result in the most efficient operation of the physical network.

MARKET POWER ISSUES

Market power has been defined as the ability of a supplier to profitably raise prices above competitive levels and maintain these prices for a significant period of time. There are two types of market power: vertical and horizontal. A traditional vertically integrated utility might exercise vertical market power by using its control of the transmission system to give its own generation preferential treatment. This concern has been a focus of FERC's in its restructuring orders. A supplier could have horizontal market power if it controlled a significant amount of the generation resources in an area; especially if the area cannot import power because of transmission constraints. This latter concern was addressed by many state regulators by requiring multiple buyers when generation was divested. Another form of market power that can involve both vertical and horizontal aspects is when a local supplier controls a significant amount of the fuel resources used for generation in an area. Experience has shown considerable problems with attempts to rely solely on market forces to prevent the exercise of market power. Both legal and illegal means have been used by some market participants to obtain large profits while increasing the overall cost of electricity.

Price Caps

Excessive market power, market manipulation, and even illegal market operations have contributed to wide variability in the price of bulk power electricity. As a result in some areas "price caps" have been established to limit maximum prices that can be charged in market operations. There are many arguments, pro and con, on the application of such price caps, some claiming they are essential to prevent market abuses, and others claiming they prohibit proper operation of the market.

STANDARD MARKET DESIGN (SMD)[4]

In July 2002, FERC issued a Notice of Proposed Rulemaking (NOPR) "Remedying Undue Discrimination Through Open Access Transmission Service and Standard Electricity Market Design", which is referred to as the SMD NOPR. It deals directly with the relationship of congestion relief including the seams issue and the operation of the wholesale energy market. FERC had determined that open-access to the transmission system still hadn't been achieved and offered examples to support their view. Many of the examples dealt with situations where utilities who were engaged in both the transmission and generation businesses acted to unfairly disadvantage competitors by restricting access to the transmission system. Additionally, FERC felt that inconsistent design and administration of short-term energy markets was adversely impacting the goal that electricity consumers pay the lowest price possible for reliable service.

Objectives and Goals

FERC stated its objectives as:

- To remedy remaining undue discrimination;
- To establish a standardized transmission service;
- To establish a wholesale electric market design.

It proposed to meet its objectives by requiring utilities to institute:

- A flexible transmission service;
- An open and transparent spot market design.

"The fundamental goal of the Standard Market Design requirements, in conjunction with the standardized transmission service, is to create "seamless" wholesale power markets that allow sellers to transact easily across transmission grid boundaries and that allow customers to receive the benefits of lower-cost and more reliable electric supply."

Proposals

The Commission, subject to some regional variations, proposed to:

[4] Much of the material in this section is excerpted directly from FERC's SMD NOPR. We cover the SMD in the money network chapter, since the SMD extends FERC's reach into the operation of the wholesale energy market.

- Exercise jurisdiction over the transmission component of bundled retail transactions;[5]
- Modify the existing pro forma transmission tariff to include a single flexible transmission service (Network Access Service) that applies consistent transmission rules for all transmission customers—wholesale, unbundled retail and bundled retail;[6]
- Provide a standard market design for wholesale electric markets.[7]

Recognizing the alleged abuses that were experienced to date, the order addressed the need to include "regulatory backstops" to protect customers against the exercise of market power. An important way that the market is not perfect is that it lacks customer demand response to price fluctuations. As prices rise dramatically, as in California, customers do not see the impacts because of other timing differences in billing or because the local utility must absorb the cost increases.

Transmission Owner's Options. Transmission owners were given three options:

1. Meet the definition of an Independent Transmission Provider (ITP).
2. Become part of an RTO.
3. Contract with an independent entity meeting the definition of an ITP to operate their transmission system.

Independent Transmission Providers (ITPs)

"An Independent Transmission Provider is any public utility that owns, controls, or operates facilities used for the transmission of electric energy in interstate commerce, and administers the day-ahead and real-time energy and ancillary services markets in connection with its provision of transmission services pursuant to the SMD Tariff, and that is independent (i.e., has no financial interest, either directly or through an affiliate, in any market participant in the region in which it provides transmission services or in neighboring regions.)"[8]

"ITPs" would be responsible for:

[5] Currently regulated by state commissions.

[6] Eliminating the dual pricing system allowed in prior orders.

[7] Recognizing that continuing to allow each area the freedom to develop its own market structure with its own rules was untenable if transactions were to be encouraged over a geographically widespread market.

[8] Recognizing the political difficulties that its proposals face and the multi-state nature of the RTOs that it envisioned, FERC indicated that State representatives would have a formal role in the ITP's decision making process. FERC avows that it does not want to interfere with the legitimate concerns of state regulatory authorities.

- The administration of the day-ahead and real-time markets for energy and ancillary services;
- Long-term planning and expansion, system impact, and facilities studies;[9] and
- Transmission transfer capability calculations (including postings on an Open Access Same-Time Information System (OASIS)).

Transmission Charges

A single transmission charge would be established. "The proposed Network Access Service would combine features of both existing open-access transmission services—the flexible and resource and load integration of Network Integration Transmission Service; and the reassignment rights of Point-to-Point Transmission Service. It would give a customer the right to transmit power between any points on the transmission system—so long as the transaction is feasible under a security constrained dispatch."

Access charges reflecting the cost of service of transmission owners would be collected on a pro-rata basis from all customers taking power off the grid and not on the generators connected to the grid. The charge could be either based on the zone of delivery within an area (a license plate rate) or could be the same for the entire ITP's area (a postage stamp rate). Based on the notion that charges to transmit energy from distant generation should reflect the actual variable transmission costs, FERC proposes that customers receiving energy from another area only pay the access charges in their own home area.

"All customers would pay congestion costs and losses associated with their particular transaction." Congestion would be managed by the use of a bid-based, security constrained spot market using locational marginal pricing (LMP).

To provide a hedge against incurring congestion costs, customers can buy congestion revenue rights. "Congestion Revenue Rights provide the rights holder with the revenues associated with congestion between the associated points; thus, any congestion costs it pays are fully offset by these revenues. To the extent the Congestion Revenue Rights holder opts not to schedule transmission service at those points, it would still receive the congestion revenues."

To give the ITP the ability to access the potential for congestion, "The customer must identify the ultimate source and sink (i.e., sources of generation and locations of customer load) so that the various system operators in an interconnection can assess the simultaneous feasibility of all scheduled power flows."

[9] These studies would include studies of the interconnection of a new load or generator, studies of the feasibility of simultaneous transactions to deal with congestion revenue rights.

Wholesale Electric Market Design

To remedy the negative impacts it perceives because of differing market structures, FERC proposed a new standardized market design. The wholesale market would rely on both:

- Bilateral contracts between buyers and sellers;
- Short-term spot markets.

FERC feels that the use of bilateral contracts will lessen the impact if power generators, presently supplying local areas, seek higher priced markets elsewhere since the local consumers could buy the power under long-term contracts.

The short term spot markets would:

- Be operated by the ITPs;
- Be bid-based, security constrained;
- Cover two time frames;
  - A day ahead
  - Real time
- Use a market-based locational marginal pricing (LMP) transmission congestion management system.

The market would have:

- Tradable financial rights to allow a fixed price for transmission service (congestion revenue rights);
  - An auction process to allocate these rights is FERC's preferred approach
- A power market monitoring system and market power mitigation rules.

FERC's efforts to establish a system of a financial congestion rights (CRR) under the SMD procedures have raised some questions, such as scheduling priorities for physical transactions, allocation of such rights, and so on. One proposal is to use the funds from such rights to help construct new transmission facilities.

Locational Marginal Pricing (LMP). The basic definition of an LMP is "the price of supplying an additional MW of load at each location (bus) in the system."[10] The major factors affecting the LMP values are the generator bid prices, the transmission system elements that are experiencing congestion, the losses on the system, and the electrical characteristics of the system.

[10] Note it does not say the cost of supplying an additional MW, it says the price.

Although there are some variations from system to system, the basic usage of LMPs is as follows: The LMP is equal to the bid price of the last, and hence most expensive, generator that is scheduled to meet the load. If there are no transmission constraints, the LMP is the same at all points on the system. If there are transmission constraints, generation is redispatched to avoid the constraint. In this situation there will be more than one LMP, since the process reflects the bid prices of different marginal generation, depending on their location in the system. Generators are paid for the energy they supply to the market, according to the LMP at their point of connection to the system. Energy consumers buy the energy they buy from the spot market, based on the LMP at their connection point. Bilateral transactions pay a congestion charge that is based on the difference in the LMPs between the delivery point of the transaction and the receipt point. Transactions into or out of the system pay congestion charges based on their entry or withdrawal point.

LMP values may be calculated for different time periods based on the particular rules for the various systems. Many of them have a day ahead market that uses the scheduled quantities of consumption for the various market players, the schedules of bilateral contracts, the price bides, and the impact of transmission congestion to determine the day ahead LMP values. The calculation of the LMP values is based on the optimization problem inherent in the market clearing process. Some systems also have an hour ahead market. Hour ahead LMP values can again be calculated as a byproduct of the hour ahead market clearing process. Finally, real-time LMP values are calculated. These are based on the generation dispatching process used for balancing the system while alleviating congestion. Differences between the LMPs are used as the cost of congestion between any two locations. Loss effects are also included in the calculations of LMPs.[11]

Some experts believe that the use of LMP increases the cost of electricity over the use of incremental production costs for dispatch and also can increase transmission losses.

Resource Adequacy

Each area would have a resource adequacy requirement, the level to be determined on a regional basis by an RTO or other regional entity including state representatives. The RTO would:

- Forecast the region's future resource needs;
- Facilitate regional determination of an adequate future level of resources;

[11] The method of calculating losses varies across regions. Some use a fixed quantity expressed in percent applied to all transactions during all hours. Other regions calculate incremental losses based on the actual transactions by using load-flow simulations.

- Access the adequacy of the plans of load-serving entities to meet the regional needs.

The obligation to meet the resource adequacy requirement would be given to the load serving entities (LSEs). "Each load-serving entity would be required to meet its share of the future regional need through a combination of generation and demand reduction. A minimum reserve requirement of 12%, well below levels employed until recently, was proposed. Additionally, all users of the transmission system would need to comply with standards for ensuring system security and reliability.

The responses to FERC's requests for comments have been varied, and in many instances heated. Some commentators question whether the proposed market will achieve FERC's goals or lead to an increase in the costs of electricity while the reliability of the system is allowed to deteriorate.

Transmission Tariffs

Recognizing that relying on financial congestion rights and the use of regulated rates of return and standard depreciation lives has not proved to be enough of an inducement to build new transmission, FERC has proposed an incentive pricing policy.[12]

Merchant Transmission

Proposals are being made for independent organizations to build "merchant transmission", that is, transmission lines which will be independently owned and on which transmission service will be provided. Major questions have been raised concerning such transmission projects, particularly their ability to fit into future long-range system needs.

Markets for Buying and Selling Rights

A key aspect of restructuring efforts on the money network is the establishment of approaches for buying and selling rights such as transmission rights and emission rights.

[12] See FERC's "Proposed Pricing Policy for Efficient Operation and Expansion of the Transmission Grid", issued January 15, 2003.

15

INFORMATION, COMMUNICATIONS AND CONTROL NETWORK

This network is an extensive system impacting the physical, energy, money, business and regulatory networks. It includes all mechanisms and processes for obtaining information, transmitting this information to some location where the information is analyzed and decisions made as necessary and then the communication of the decisions so they can be acted upon.

This network is not static in time. We've discussed some of the technology improvements and innovations being implemented in the physical network. Similarly, the information, communications and control networks is also experiencing similar changes.

The utility industry has been affected, as have many others, by the widespread use of the internet, communication networks capable of transmitting large quantities of information very quickly, and computer based sensing and computing equipment based on microprocessor technology. These developments have allowed improvements in many areas of the business; in reliability, in productivity, in speed of response, and so forth.

One aspect of the development is the improvements made in so called "legacy" computer systems internal to the utilities. In many cases, these systems were characterized by stand alone data bases containing information that, if properly combined, would be of great benefit to the company.

In the following material we will mention some, but by no means all, of the components of this network.

Understanding Electric Power Systems: An Overview of the Technology and the Marketplace, by Jack Casazza and Frank Delea
ISBN 0-471-44652-1

FINANCIAL AND BUSINESS OPERATIONS

The introduction of advanced financial management systems allow managers to better understand the cost drivers of their business and business segments and to deal with the increasing complexity of the utility business structure, especially on the holding company level. For example, interrelationships of electric trading strategies, purchasing decisions, outsourcing strategies can all be examined.

Computer technology and the internet have had a major impact on the way companies do business. The terms e-Business and e-Commerce are used to describe a wide spectrum of applications that allow the utility to exchange information with its customers, its suppliers, its regulators and the energy market:

- The internet based OASIS System is at the heart of the wholesale energy market.
- Internet-based material and services procurement arrangements with suppliers eliminate processes that added to the time and cost of obtaining these services.
- Customers can now pay their bills online.
- Customer contact has been improved by the establishment of customer call centers operated 24 hours a day.
- Online filing of documents with the SEC and other regulatory agencies in now in operation.

The OASIS System is a good example of the interconnectivity of the various networks. For example, it deals with the physical security of the transmission system as well as the financial system since it deals with the submission of bids for various generation related products and services.

SYSTEM OPERATIONS

A measurement and communication network has been used for years by the industry to measure and transmit, to a central control point, information about the physical state of the power system (voltages, currents, breaker and relay status, generator output, etc). The network is also used to transmit operational orders to substations and to generating stations:

- In recent years high-speed communication links and online state estimation[1] have been implemented.

[1] The use of tele-metered values to rapidly determine the electrical status at all points on the transmission system and to suggest remedial actions in case of emergencies.

- Computer programs are being developed to dynamically analyze the optimum settings of FACTS devices to increase the reliability and power transfer capability of the transmission system on a real time basis.

DISTRIBUTION OPERATIONS

Significant strides have been made in the ability to automate the operation and reduce the costs associated with the distribution system. Many of the benefits were listed in Chapter 7. Among these are:

- Work management systems (WMS) used for optimizing work scheduling and dispatching of work orders to field crews;[2]
- Improvements, using microprocessors, in SCADA systems for remote monitoring and control of the status of the distribution system; that is, line loadings, circuit breaker status, voltages to a central operations center;
- Geographic information systems (GIS),[3] which relate customers to specific geographic locations;
- Outage-management systems (OMS), to handle the large amount of information relating to system outages, especially outages impacting many customers due to ice storms, tornadoes, etc. and the efficient dispatching of crews for restoration purposes;
- Asset-management systems which are used to schedule maintenance (AMS);
- Automated meter reading (AMR) which offers the potential for significant reduction in the labor intensive present system.

Some utilities are combining aspects of these systems to create a more robust tool. For example, integrating a GIS system with an OMS system, and using information from 24-hour call centers and from the customer information system (CIS) provides additional information which can shorten restoration times.

[2] Information on WMS can be found in "Utility Automation". See the January 2003 issue for a broader discussion of issues concerning WFM in an article by Scott Munro, "Work Force Management Unifies Field Operations".

[3] Information on GIS systems can be found at the USGS Website: www.usgs.gov. The following was excerpted from that site: "In the strictest sense, a GIS is a computer system capable of assembling, storing, manipulating, and displaying geographically referenced information, i.e. data identified according to their locations. Practitioners also regard the total GIS as including operating personnel and the data that go into the system." "The way maps and other data have been stored or filed as layers of information in a GIS makes it possible to perform complex analyses."

PHYSICAL SECURITY

Issues dealing with the physical security of the bulk power system have taken on a new meaning since the attack on September 11, 2001.

The following material was taken form NERC's website, where the reader can find additional information.

The effort, on a national level, to deal with electric utility security began in the late 1990s with a Presidential Order "Protecting America's Critical Infrastructures." Electric utilities were one of eight areas identified as critical.

In response to the Order, the DOE was appointed the lead agency for the electric sector. The DOE appointed NERC as the Sector Coordinator, "to:

- Assess sector vulnerabilities;
- Develop a plan to reduce electric system vulnerabilities;
- Propose a system for identifying and averting attacks;
- Develop a plan to alert electricity sector participants and appropriate government agencies that an attack is imminent or in progress;
- Assist in reconstituting minimum essential electric system capabilities in the aftermath of an attack."

The industry responded by addressing the following areas:

- Deterrence;
- Detection;
- Assessment;
- Communications;
- Response critical."

As discussed earlier, the electric system is designed to withstand various disturbances. The industry has also maintained the capability to rapidly restore service in the event of outages. In addition to the physical assets, attention is also focused on the SCADA systems used for control and communications. Disruption of this system can have wide-ranging impacts, not only on the physical performance of the system (bogus telemetered orders to open many circuit breakers at the same time resulting in the loss of more facilities than considered in design), but also on the commercial operation of the system.

COMMERCIAL SECURITY

Commercial security has two aspects:

1. The ability of the market to avoid manipulation by exercise of market power or by individuals who illegally violate market rules;

2. The ability of the market to withstand disruption or manipulation by individuals who illegally gain access to its communication or computer systems.

FERC has focused its attention on the first aspect ever since it initiated its efforts on restructuring the industry. Experience in California has shown that great sums of money can be made by exercising various forms of market power or by violating market rules. It is for this reason that FERC proposed a market monitoring function in its SMD NOPR.

The second aspect is a one of the focuses of the nation's response to terrorist threats. NERC has efforts under way in this area.

Cyber attacks can be of two types:

1. Those against an individual or specific functions, such as theft of identification and credit card numbers of an individual;
2. Those against an entire system, such as viruses which attack entire internet systems or against an individual company and all of its files.

As companies rely more and more on computerizing and networking their systems, they also make themselves more vulnerable to cyber attack. Efforts to standardize and to use open platforms contribute to this vulnerability. The efforts to combine data base information discussed in the Distribution Operations section above, illustrate the increased exposure when combinations are implemented.

The literature contains many articles by specialists who address issues such as protocols for access to sensitive information and the technology of building robust firewalls.[4]

[4] See the "National Strategy to Secure Cyberspace" issued by the White House in February 2003, at www.whitehouse.gov/pcipb/cyberspace_strategy.pdf

16

ROLE OF NERC, NAESB AND OTHER ORGANIZATIONS

The roots of the North American Reliability Council (NERC) lie in the Northeast blackout of 1965 and the PJM blackout of 1967. It became generally recognized that measures were required that would:

- Ensure that adjacent systems would be planned, designed, and operated based on consistent reliability criteria; and
- Establish appropriate criteria.

To achieve these goals, various regional reliability councils were formed and began functioning. It soon became apparent that a national organization was required to ensure coordination between the various regional councils. While some differences in reliability standards and operating procedures were feasible, if adjacent regions were not affected adversely by them, others were not. This lead to the formation of what is now called the North American Electric Reliability Council (NERC).[1] The various regional councils were the members of NERC, with NERC activities almost entirely controlled by the utility systems that were members of these regional councils.

As restructuring advanced, membership in the reliability councils changed to include all segments of the industry, including independent generators, power marketers, and so forth. As a result, a review of the overall NERC struc-

[1] This originally was the National Electric Reliability Council, but the name was changed when Canada and Mexico became involved.

Understanding Electric Power Systems: An Overview of the Technology and the Marketplace, by Jack Casazza and Frank Delea
ISBN 0-471-44652-1

ture was made and plans for a revised reliability organization developed, to be called the North American Electric Reliability Organization (NAERO).

The revised organization provided for NERC to establish reliability standards, monitor compliance, and establish penalties for non-compliance. A major decision of NERC/NAERO however, was that it would deal only with reliability matters and not become involved with business, or related monetary disputes.

NERC, Reliability Councils, and RTOs

A significant problem remains in the functioning of the reliability councils and NERC. As new RTOs are formed, their membership is not congruent with the existing reliability councils. In some cases an existing reliability council may cover portions of two RTOs. In other cases several reliability councils, possibly having differing reliability standards, may be in a single RTO.

NAESB[2]

For some years business standards in the gas industry were established by a gas standards organization. Discussions with NERC lead to this organization becoming the North American Energy Standard Board (NAESB) which would establish business standards for the electric power industry. The chairman of NERC and NAESB have signed a memorandum of understanding (MOU) designed to ensure that the development of wholesale electric business practices and reliability standards are harmonized and that every effort is made to eliminate duplication of efforts between NERC and NAESB.

The MOU calls for regular communications between the two organizations about their respective standards development activities, including a requirement that NERC and NAESB review and comment on each other's annual work plans. The MOU also establishes a Joint Interface Committee ("JIC") that will review all standards development proposals received by either organization to determine whether NERC or NAESB should draft a particular standard.

NAESB is presently contemplating having the various industry sectors propose such business standards. They will assist in arriving at a consensus which will become the official industry standards.

Enforcement and Dispute Resolution

As both enforceable reliability and business standards are established with significant penalties for non-compliance, it become generally recognized that an increased number of disputes would arise. Disputes have been frequent in

[2] The material in this section is based on NAESB's Website: www.naesb.org

the past, many of which have resulted in FERC hearings, and sometimes court challenges, involving considerable time and expenditures. FERC has established a dispute resolution department with the responsibility for encouraging various types of alternative disputes resolution procedures to reduce the time and costs required for resolution. These simple procedures would require many more mediators and arbitrators. Educational programs to train these mediators and arbitrators are becoming more important.

PROFESSIONAL ORGANIZATIONS

Many professional organizations are involved in the functioning of the electric power industry.

IEEE[3]

"The Institute of Electrical and Electronics Engineers, Inc. (IEEE) is a non-profit, technical professional association of more than 377,000 individual members in 150 countries. Through its members, the IEEE is a leading authority in technical areas ranging from computer engineering, biomedical technology, and telecommunications, to electric power, aerospace, and consumer electronics, among others. The IEEE is made up of:

- 10 regions;
- 37 societies;
- 4 councils;
- Approximately 1,200 individual and joint society chapters;
- 300 sections; and
- 1,000 student branches are located at colleges and universities worldwide."

The Power Engineering Society is one of the 37 societies in the IEEE and has 25,000 members.

"Through its technical publishing, conferences, and consensus-based standards activities, the IEEE:

- Produces 30 percent of the world's published literature in electrical engineering, computers, and control technology;
- Holds annually more than 300 major conferences; and
- Has nearly 900 active standards with 700 under development."

[3] The material in this section is from the IEEE/PES website: www.ieee.org/organizations/society/power

"Policy matters related to IEEE Standards are the purview of the IEEE Standards Association (IEEE-SA), which establishes and dictates rules for preparation and approval . . . Overwhelmingly, it is the Computer Society and the Power Engineering Society that dominate in this regard, for instance, about 40% of all IEEE Standards are . . . within the PES."

CIGRE

Another important organization is the International Council on Large High Voltage Electric Systems (CIGRE). CIGRE is an international organization through which ideas can be exchanged with people from various countries through meetings, committee activities, and its publications.[4]

INDUSTRY ASSOCIATIONS

To exchange ideas and to provide an effective voice in government and public relations circles, a number of business organizations have been formed.

NARUC[5]

"The National Association of Regulatory Utility Commissioners (NARUC) is a non-profit organization founded in 1889. Its members include the governmental agencies that are engaged in the regulation of utilities and carries in the fifty states, the District of Columbia, Puerto Rico, and the Virgin Islands. NARUC's member agencies regulate the activities of telecommunications, energy, and water utilities.

NARUC's mission is to serve the public interest by improving the quality and effectiveness of public utility regulation. Under state law, NARUC's members have the obligation to ensure the establishment and maintenance of such energy utility services as may be required by the public convenience and necessity, and to ensure that such services are provided at rates and conditions that are just, reasonable, and nondiscriminatory for all consumers."

AEIC[6]

"The Association of Edison Illuminating Companies (AEIC) is an association of electric utilities, generating companies, transmitting companies and distributing companies in North America and oversees. Organized in 1885, the AEIC is the oldest association to be affiliated with the electric utility industry. It provides information exchange through a committee structure, and mutual

[4] The material in this section is from the CIGRE website: www.CIGRE.org
[5] The material in this section is from NARUC's website: www.naruc.org
[6] The material in this section is from the AEIC's website: www.aeic.org

solutions to industry problems, as well as providing literature on load research and underground cable specifications."

"The purpose of the Association of Edison Illuminating Companies is:

- To promote the (technology related) business interest of its members;
- To discover and adopt increasingly more reliable, economical, and efficient means for the supply and utilization of electrical energy,
- To provide an assembly for the exchange of experiences of electrical properties."

APPA[7]

"The American Public Power Association (APPA) is the service organization for the nation's more than 2,000 community-owned electric utilities that serve more than 40 million Americans. It was created in 1940 as a non-profit, non-partisan organization. Its purpose is to advance the public policy interests of its members and their consumers, and provide member services to ensure adequate, reliable electricity at a reasonable price with the proper protection of the environment."

"APPA is governed by a regionally representative Board of Directors."

EEI[8]

"Edison Electric Institute (EEI) organized in 1933, is the Association of United States shareholder-owned electric companies, international affiliates and industry associates worldwide. In 2000, its U.S. members served more than 90% of the ultimate customers in the shareholder-owned segment of the industry, and nearly 70% of all electric utility ultimate customers in the nation. They generated almost 70% of the electricity generated by U.S. electric utilities."

EEI's mission is to ensure members' success in a new competitive environment by:

- Advocating public policy;
- Expanding market opportunities;
- Providing strategic business information.

"EEI works closely with its members, representing their interests and advocating equitable policies in legislative and regulatory arenas. . . . the Institute provides authoritative analysis and critical industry data to its members, Congress, governmental agencies, the financial community, and other influen-

[7] The material in this section is from APPA's website: www.appanet.org
[8] The material in this section is from the EEI's website: www.eei.org

tial audiences. EEI provides forums for member company representatives to discuss issues and strategies to advance the industry and to ensure a competitive position in a changing marketplace."

ELCON[9]

The Electricity Consumer Resource Council (ELCON), founded in 1976, is "an association of large industrial consumers of electricity . . . coming from virtually every manufacturing industry. They consume nearly six percent of all the electricity used in the United States." ELCON describes itself as "a . . . voice for competitive policies and market structures." It strives to lower electricity costs for its industrial members.

NRECA[10]

"The National Rural Electric Cooperative Association (NRECA) is the national service organization dedicated to representing the national interests of consumer-owned cooperative electric utilities and the consumers they serve. The association provides . . . legislative representation before the U.S. Congress and the Executive Branch, and representation in legal and regulatory proceedings affecting electric service and the environment.

NRECA also offers educational and training programs for cooperative directors, managers, and employees; collaborative research to enhance cooperatives' use of technology and insurance, employee benefits and financial services and technical advice and electrification assistance in developing countries around the world."

NRECA's electric cooperative and public power district members serve 36 million people in 47 states. Approximately 855 NRECA members are electric distribution systems. NRECA membership includes other organizations formed by these local utilities:

- 64 generation and transmission cooperatives for power supply;
- Statewide and regional trade and service associations;
- Supply and manufacturing cooperatives;
- Data processing cooperatives;
- Employee credit unions.

Associate membership is open to equipment manufacturers and distributors, wholesalers, consultants, and other entities that do business with members of the electric cooperative network."

[9] The material in this section is from ELCON's website: www.elcon.org
[10] The material in this section is from NRECA's website: www.nreca.org

Electric Power Supply Association[11]

"EPSA is the national trade association representing competitive power suppliers, including independent power producers, merchant generators and power marketers. These suppliers, who account for more than a third of the nation's installed generating capacity, provide reliable and competitively priced electricity from environmentally responsible facilities serving global power markets. EPSA seeks to bring the benefits of competition to all power customers. Formed as a result of a merger between the National Independent Energy Producers and the Electric Generation Association, EPSA combines the strengths and policy successes of those two prominent organizations on behalf of the competitive power supply industry. EPSA's formation has given the competitive power supply industry the ability to speak with a single, unified voice at the global and national levels."

RESEARCH ORGANIZATIONS

Years ago two types of research organizations were founded. One concerned with technical research and one with regulatory research.[12]

EPRI[13]

"EPRI, the Electric Power Research Institute, was founded in 1973 as a non-profit energy research consortium for the benefit of utility members, their customers, and society. Its mission is to provide science and technology-based solutions . . . by managing a far-reaching program of scientific research, technology development, and product implementation. . . . Serving the entire energy industry—from energy conversion to end-use in every region of the world."

"EPRI's technical program consists of:

- Power generation;
- Distributed resources;
- Nuclear;
- Environment;
- Power delivery and markets."

"EPRI's programs are open to all organizations involved in the energy industry, encompassing:

[11] Information obtained from http://www.epsa.org

[12] Both EPRI and NRRI were formed at the suggestion of Joseph Swidler, former Chairman of the Federal Power Commission.

[13] The material in this section is from EPRI's website: www.EPRI.com

- All power utilities, including investor-owned, municipalities, cooperatives, and federal government utilities;
- Competitive power producers, energy service companies, engineering service companies, natural gas entities, power marketers, manufacturers, industrial companies, and other energy suppliers;
- Independent system operators, power exchanges, power scheduling coordinators, transmission companies, distribution companies, and nuclear licensees;
- Government organizations involved in funding public-benefit R&D programs."

Other Research[14]

Industry-related research is also conducted under the auspices of the DOE, the Empire State Electric Energy Research Corp. (ESEERO), and by numerous universities funded by both the industry and the Federal Government.

NRRI[15]

The National Regulatory Research Institute (NRRI) was established by the National Association of Regulatory Utility Commissioners at Ohio State University in 1976. It is the official research arm of NARUC. It provides "... research and services to inform and advance regulatory policy, primarily for U.S. state public utility commissions."

"The NRRI has four programs of regulatory research and assistance: infrastructure, markets, consumers, and commissions. The programs address regulatory policy in the electricity, natural gas, telecommunications, and water industries.

[14] Links to many other research organizations can be found at www.netl.doe.gov/weblinks/links.html

[15] The material in this section is from NRRI's website: www.nrri.ohio-state.edu

17

WHERE RESTRUCTURING STANDS

Electricity costs have increased significantly from what they might have been if the former industry structure and regulatory procedures had remained in place.[1] Claims are made that this will be corrected once market abuses are eliminated and effective market operations are established, but many experts have their doubts. Some new studies show that, because of its nature, the electricity business results in lower costs to consumers and higher levels of reliability with cooperative rather than the new competitive procedures.

To justify the new regulatory approach, claims have been made that it will result in lower costs to consumers. These claims seemed to be based on a belief that the regulatory regime results in higher costs because of mismanagement and the propensity to overbuild in order to increase rate base and, therefore, profits.

From a quantitative perspective, these claims have not been supported. However, from a qualitative perspective, numerous pressures can be identified for increased costs to consumers under the new regulatory regime. Some of the factors that exert upward pressure on the costs to consumers are:[2]

[1] Transmission Access and Retail Wheeling: The Key Questions, pp. 77–102, in *Electricity Transmission Pricing and Technology*, an EPRI book, Kluwer Academic Publishers, 1996.
[2] No attempt has been made to present these factors in an order that relates their impact on costs to the consumer.

Understanding Electric Power Systems: An Overview of the Technology and the Marketplace, by Jack Casazza and Frank Delea
ISBN 0-471-44652-1

- A higher risk premium required by generating companies on their investments in the new environment because of an increase in uncertainty;
- For those companies that purchased existing power plants at multiples of their book value, the higher returns needed to recoup their investment;
- Shorter economic service lives and hence higher annual depreciation expenses for generators;
- An increase in the total amount of generating capacity than would otherwise have been necessary to have competitive generation markets work effectively;
- The use of LMP pricing for dispatched generation where the price of the highest bid generation needed to supply the customer requirements sets the price to all customers, rather than the average cost of all generation run as in the prior situation;
- The recently announced FERC proposal to allow a risk premium of up to 200 basis points to owners of new transmission;
- The increased complexity and, therefore, costs of managing the transmission system;
- The increased number of individuals required by the various entities to do the same functions as were handled by a single staff in a regulated company.

Offsetting these upward pressures are assumptions that:

- Merchant power plants will be installed at a significantly lower capital cost per megawatt than in the past[3];
- Power plants will be better maintained thus increasing their availability and, thereby, lower the total amount of generation capacity needed[4]

Along with these effects has been a significant amount of illegal and unethical[5] actions by those involved in the electric power industry, including fraudulent accounting and "gaming procedures" to increase profits.

[3] To some extent, this assumption ignores the reality that in some areas of the country building of new generation continues to be extremely time consuming and, hence, costly because of stringent siting requirements by local governments.

[4] A lower reserve margin does not necessarily result in lower consumer costs, since a lower reserve can result in higher price generation being at the margin and, therefore, setting the LMP for many more hours of the year.

[5] *Ethics and Profits—The Crisis of Confidence in American Business*, Leonard Silk and David Vogel, Simon and Schuster, 1976.

REQUIRED ADDITIONAL ANALYSES

An increasing number of engineers, universities, and other organizations have been asking for analyses of the past results of deregulation and of proposed future policies. While price comparisons have been made, some showing reductions from prior prices and others showing increases, these are strongly influenced by other cost changes, that is, fuel, the cost of money, subsidies, taxes, and so on, but the studies are generally not meaningful. Analyses based on the effects on costs are now being proposed. Only if costs have been decreased from what they would otherwise have been can prices become lower or profits higher because of deregulation.

The advocates of such cost analyses stress that no sound business would be run based on policies whose results are not evaluated from time-to-time. They claim that the continuing focus on "market forces" to develop future policies without any analyses of the effect of these market forces on the costs of electricity perpetuates past errors. The continuing failure to make evaluations of the results of restructuring is hard to understand and is raising concerns that some are afraid of the answers. Only by thorough cost analyses can the benefits and harm of restructuring be determined thus enabling policies to be established retaining what is good and discarding what is bad.

ABANDONMENT OF DEREGULATION

Some states that implemented deregulation are deciding that it was a mistake and are abandoning it, reverting to the previous regulatory structure. These include California and the Province of Ontario in Canada. Other states are postponing steps to deregulate, and others, including Florida, have decided against deregulation.

POWER SUPPLY

2002

Electric power supplies were generally adequate in 2002, a year in which weather was mild, electricity use down, and significant power plant additions were made. In 2002 more than 45,000 MW of new generation was completed. Nearly all of these plants were supplied by natural gas. This 7% growth in new supply was adequate with the increase in electricity demand being only 2.8%.

The Future

The disappearance in the huge profit incentives to build new power plants has caused a major decline in the building of new power plants, with more than

100,000 MW of new power plants being cancelled or postponed during 2002. Adding to the decreased incentives to build power plants has been the huge debts accumulated by existing plant owners. About $25 billion of this debt is coming due in 2003. This is leading to a downgrading of the credit ratings of almost half of the utility holding companies and merchant plant owners. This increases their future costs of money for both equity and debt.[6]

Energy Trading. Along with the financial difficulties of the owners and builders of the power plants has been the sharp decline in the number and scope of activities of energy trading companies established by utilities and independent trading organizations as a result of large financial losses and a number of illegal activities.

RELIABILITY CONCERNS

Pressure to maintain a lower installed generation reserve increases the risks that some customer load can't be supplied and the inability of the competitive market to work effectively. Considerable concern exists about the ability to meet our future national electric power requirements reliably and economically. Some believe that cancellation of major generation projects could lead to political pressures for government takeovers of our electric power systems.

Transmission Problems

The difficulties in building the additional transmission lines needed at the present time and in the future continue to exist. State regulatory authorities conduct extensive hearings on plans to build new lines. These hearing require long-range justification for the lines which has become increasingly difficult to provide since the future generation locations that determine transmission requirements are not known, even for a few years in the future. Needed lines are being delayed as a result.

Investment in new transmission, expenditures for reinforcement, and maintenance of existing facilities by the transmission owners has declined significantly since earnings on such investment are difficult to recover. The FERC regulation of both the allowable cost recovery period (depreciation) and the return on investments is considered too low and has been discouraging the needed improvements.

NATIONAL POWER SURVEY

From the early 1920s, it has been the practice in the United States for the federal government to sponsor "national power surveys" about every 10 or 15

[6] A number of merchant power plant owners have sought bankruptcy protection.

years. These surveys have been conducted cooperatively by the government, the power industry, and nationally known engineers. These surveys reviewed future long-range generation and transmission needs, considering potential future scenarios. While not a blueprint for the future, they provided guidance to those developing power systems, planning future technology developments, and sponsoring research.

Pressures for a new "national power survey" continue to mount. While each of the regions performs studies to determine future transmission needs, usually under ISO or RTO sponsorship, a national power survey would provide a mechanism to coordinate as a national basis these regional plans. Such questions as the use of DC transmission, the best transmission voltage to use, which circuits offered the best opportunities for inter-regional transfers, and the effect of one region's plans on another regions could all be reviewed.

CONCLUSIONS

A number of possible developments for the future have been outlined above. Decisions on these problems and procedures will take time. While this is happening, two additional steps appear likely:

1. Increased use of "performance base" rate making;
2. Development of "coordination contracts".

Performance based ratemaking on which rates are based on overall performance and customer's satisfaction instead of purely financial data will overcome some of the inadequacies of former rate making procedures. It will increase to the rewards for good management.

With the unbundled industry as it now exists, significant benefits are achievable to both the market participants and the consuming public from "coordination contracts". Coordination contracts provide a means of sharing costs and benefits in developing plans and operating to minimize total electricity costs. These contracts would cover arrangements under which a participant could take actions involving increased costs for itself that benefited others by a greater amount, thus creating overall system savings. These contracts would be negotiated between the parties to determine cost allocation so that all participants benefit and all get an appropriate share of the savings. They are a way to restore some of the benefits of coordination in a competitive world.

Competition in the power supply market continues to be developed. Various marketing procedures are still under consideration, varying from FERC's proposed SMD to the development of "regional energy service commissions", through which groups of states could select their own procedures in setting up their regional markets.

Retail wheeling and customer choice faces a far more negative future. Evidence is increasing that it will not develop in the future, and may be abandoned where it exists.

In general, the pace of new legislation and regulation may slow as a result of the continuing calls for a "time-out" to evaluate the future alternatives.

INDEX

Understanding Electric Power Systems: An Overview of the Technology and the Marketplace, by Jack Casazza and Frank Delea
ISBN 0-471-44652-1

ABOUT THE AUTHORS

Jack Casazza is presently President of the American Education Institute and an Outside Director of the Georgia Systems Operation Company. Formerly a corporate officer for a large utility in the United States and an executive with major consulting firms, he has been responsible for forecasting; DSM studies; integrated system planning; developing generation and transmission plans; economic and financial evaluations; intercompany contract negotiations; rate and cost of service studies; merger studies and analyses; strategic planning; organizational planning; pooling and coordination studies; system reliability standard establishment and evaluations; cogeneration analysis; research programs; and applications of new technology. He has been responsible for consulting projects in the USA, South America, Canada, Africa, Asia, Australia, and Europe. He is an IEEE Life Fellow.

Frank Delea retired in 1997 from Consolidated Edison Company, where he had a wide range of assignments giving him insight into planning, operational, financial, organizational and legal issues. He was the Company's Chief Electric Planning Engineer, its Chief Forecast Engineer, and its first Project Manager for Rate Cases. As Director of Corporate Planning, he was involved in many issues faced by the evolving utility industry including mergers and acquisitions, investments in non-regulated subsidiaries and corporate restructuring. He was leader of a team charged with the formation of a holding company. He also was a member of the Generation and the Financial Teams charged with developing business strategies for a non-regulated generation subsidiary and for the sale of regulated generation assets. He is presently associated with the American Education Institute where he conducts short courses in technical, business, and regulatory issues relating to electric power systems for non-engineers. He is a member of the IEEE.

Understanding Electric Power Systems: An Overview of the Technology and the Marketplace, by Jack Casazza and Frank Delea
ISBN 0-471-44652-1

Printed in the United States
79485LV00002B/130-249